ASCE STANDARD

ASCE/COPRI

77-22

AF767524

Dry Dock Standard

ASCE
AMERICAN SOCIETY OF CIVIL ENGINEERS

COPRI
ASCE

PUBLISHED BY THE AMERICAN SOCIETY OF CIVIL ENGINEERS

Library of Congress Cataloging-in-Publication Data

Names: American Society of Civil Engineers, author.
Title: Dry dock standard.
Description: Reston, Virginia : American Society of Civil Engineers, [2023] | Series: ASCE standard | Includes bibliographical references and index. | Summary: "Dry Dock Standard, ASCE/COPRI 77-22, provides guidance on safety certification of commercial drydocking facilities through maintenance and inspection that will minimize risk to personnel and ships" – Provided by publisher.
Identifiers: LCCN 2023000707 | ISBN 9780784416006 (soft cover) | ISBN 9780784483923 (pdf)
Subjects: LCSH: Dry docks–Maintenance and repair–Standards. | Dry docks–Safety measures–Standards.
Classification: LCC TC361 .A474 2023 | DDC 623.8/3–dc23/eng/20230130
LC record available at https://lccn.loc.gov/2023000707

Published by American Society of Civil Engineers
1801 Alexander Bell Drive
Reston, Virginia, 20191-4382
www.asce.org/bookstore | ascelibrary.org

This standard was developed by a consensus standards development process that has been accredited by the American National Standards Institute (ANSI). Accreditation by ANSI, a voluntary accreditation body representing public and private sector standards development organizations in the United States and abroad, signifies that the standards development process used by ASCE has met the ANSI requirements for openness, balance, consensus, and due process.

While ASCE's process is designed to promote standards that reflect a fair and reasoned consensus among all interested participants, while preserving the public health, safety, and welfare that is paramount to its mission, it has not made an independent assessment of and does not warrant the accuracy, completeness, suitability, or utility of any information, apparatus, product, or process discussed herein. ASCE does not intend, nor should anyone interpret, ASCE's standards to replace the sound judgment of a competent professional, having knowledge and experience in the appropriate field(s) of practice, nor to substitute for the standard of care required of such professionals in interpreting and applying the contents of this standard.

ASCE has no authority to enforce compliance with its standards and does not undertake to certify products for compliance or to render any professional services to any person or entity.

ASCE disclaims any and all liability for any personal injury, property damage, financial loss, or other damages of any nature whatsoever, including without limitation any direct, indirect, special, exemplary, or consequential damages, resulting from any person's use of, or reliance on, this standard. Any individual who relies on this standard assumes full responsibility for such use.

ASCE and American Society of Civil Engineers—Registered in US Patent and Trademark Office.

Photocopies and permissions. Permission to photocopy or reproduce material from ASCE publications can be requested by sending an email to permissions@asce.org or by locating a title in ASCE's Civil Engineering Database (https://cedb.asce.org) or ASCE Library (http://ascelibrary.org) and using the "Permissions" link.

***Errata:* Errata, if any, can be found at https://dx.doi.org/10.1061/9780784416006.**

Copyright © 2023 by the American Society of Civil Engineers.
All Rights Reserved.
ISBN 978-0-7844-1600-6 (soft cover)
ISBN 978-0-7844-8392-3 (PDF)

Manufactured in the United States of America.
27 26 25 24 23 1 2 3 4 5

ASCE STANDARDS

In 2016, the Board of Direction approved revisions to the ASCE Rules for Standards Committees to govern the writing and maintenance of standards developed by ASCE. All such standards are developed by a consensus standards process managed by the ASCE Codes and Standards Committee (CSC). The consensus process includes balloting by a balanced standards committee and reviewing during a public comment period. All standards are updated or reaffirmed by the same process every five to ten years. Requests for formal interpretations shall be processed in accordance with Section 7 of ASCE Rules for Standards Committees which are available at www.asce.org. Errata, addenda, supplements, and interpretations, if any, for this standard can also be found at www.asce.org.

The provisions of this document are written in mandatory language and, as such, are intended to be suitable for regulatory or contractual purposes.

This standard has been prepared in accordance with recognized engineering principles and should not be used without the user's competent knowledge for a given application. The publication of this standard by ASCE is not intended to warrant that the information contained therein is suitable for any general or specific use, and ASCE takes no position respecting the validity of patent rights. The user is advised that the determination of patent rights or risk of infringement is entirely their own responsibility.

A complete list of currently available standards is available in the ASCE Library (https://ascelibrary.org/page/books/s-standards).

CONTENTS

PREFACE

This initial edition of *Dry Dock Standard*, ASCE/COPRI 77-22, is intended to address a deficiency in commercial shipyards, specifically safety certification of drydocking facilities. The US Navy has a standard (MIL-STD-1625) for certifying facilities that drydock their ships. The US Coast Guard has a procedure (SFLC 8634) for certifying facilities that drydock their ships. However, there is no commercial standard for certifying dry docks. Many dry dock owners/operators obtain "commercial certifications" from third-party engineers; however, these certifications are not performed to a standard and are typically focused on the structural capacity.

The dry dock standard provides requirements for issuing a certification that encompasses the structural design, material condition, maintenance practices, and operating procedures. This standard is intended for use by dry dock owners, Dockmasters, dry dock maintenance engineers, engineers engaged in dry dock inspection and certification, ship owners, and port engineers.

There have been many dry dock accidents over the past two decades that are attributable to inadequate maintenance and inspections. The intent of this standard is to minimize the risk to personnel and the ships being drydocked. In addition, emphasis on maintenance and inspection will provide the insight for facility owners/operators to enable repairs before the catastrophic loss of their capital asset.

ACKNOWLEDGMENTS

ASCE greatly acknowledges the devoted efforts of the Dry Dock Standards Committee, which is a subcommittee of the Ports & Harbors Committee of the Coasts, Oceans, Ports, and Rivers Institute (COPRI). This committee comprises individuals from many backgrounds. This standard was prepared through ASCE's consensus standardization process in compliance with the ASCE Rules for Standards Committees and the procedures of the ASCE's Codes and Standards Committee (CSC). ASCE's standardization process is accredited by the American National Standards Institute (ANSI).

Voting Members
Paul A. Harren, A.M.ASCE, *Chair*
John L. Watts, P.E., M.ASCE, *Vice Chair*
John R. Anderson, *Secretary*
Robert A. Ernsting, Ph.D, P.E.
Graeme Forsyth, M.ASCE
Paul R. Goetz, Retired P.E.
Robert E. Heger, P.E.
Bradley G. Lamkin, P.E.
James A. Neville
Douglas L. Pearlson, P.E.
Paul Rossetti
Scott C. Sampson
Rusty Sandidge
Waleed Sayed, P.E.
Joe Stiglich
Charles Trobaugh

Associate Members
Matt McCarty, P.E., S.E., M.ASCE
Mike Preece, P.E.
Mark A. Procter, P.E.
David W. Pryor, P.E., M.ASCE
Sieu Quan
David Von Schmidt

CHAPTER 1
GENERAL

1.1 SCOPE

This dry dock standard, herein referred to as the "standard," defines the minimum requirements to obtain and sustain dry dock certification. This standard covers the evaluation, inspection, maintenance, and operation of drydocking facilities.

To obtain dry dock certification in accordance with this standard, each drydocking facility shall undergo a *condition assessment*. On completion of the condition assessment, a dry dock certification, in the form of a Dry Dock Capacity Certificate, shall be issued for the dry dock's certified rated capacity and stability limitations (as applicable) by the certifying agency who conducted the condition assessment.

To maintain dry dock certification in accordance with this standard, the shipyard shall

- Establish and follow a maintenance program,
- Operate the facility in a safe manner as defined by this standard,
- Maintain records, and
- Have a certifying agency conduct periodic certification inspections.

Prior to the expiration date of the certification period, the shipyard shall have a dry dock certifying agency conduct *certification inspections* of the drydocking facility. On completion of the certification inspection, the certifying agency will issue an updated Dry Dock Capacity Certificate for the dry dock's certified rated capacity and stability limitations (as applicable). If the certification inspection discovers deficiencies that affect the dry dock's capacity, operational limitations, stability limitations (as applicable), or safe operation, the certifying agency shall require a new condition assessment for the deficient items.

This standard does not cover the following:

1. Personnel safety requirements;
2. Shipyard health and safety requirements;
3. Environmental and pollution compliance and control systems;
4. Design and construction of drydocking facilities;
5. Vessel handling equipment such as lines;
6. Vessel transfer systems;
7. Weight handling systems such as cranes;
8. Service systems such as potable water, heating, and ventilation; and
9. Industrial systems such as welding, burning equipment, and abrasive blast equipment.

Section 1.3 contains definitions, symbols, and notations used in this standard. Appendix A contains sample dry dock certification inspection checklists. Appendix B contains sample dry dock maintenance tasks. Appendix C contains sample drydocking checklists for use before, during, and after a drydocking operation.

1.2 APPLICATION

This standard applies to floating dry docks, graving docks, marine railways, and vertical lifts.

1.3 DEFINITIONS, SYMBOLS, AND NOTATION

Explanations of basic terms used in this standard.

Anode: An electrode at which oxidation of its surface is occurring. *See* **Cathodic Protection**.

Aspect Ratio: The ratio of a block's height divided by its base dimensions.

Ballast: Liquid or solid mass loaded on a vessel, dry dock closure gate, or floating dry dock to improve stability and trim. Temporary ballast is usually water stored in dedicated tanks. Permanent ballast is usually solid lead castings or concrete.

Baseline Height: A horizontal plane of the ship used as a reference for vertical measurement. Usually, the top of the keel blocks and identified on the docking plan.

Block Load: The force on a keel or side block owing to gravity, seismic, and wind forces acting on a vessel.

Bollard: A mooring bit, usually located alongside the top of a dry dock wall.

Bulkhead: (1) A vertical *structural* partition dividing a vessel's interior into various compartments for strength and safety purposes, (2) term applied to vertical partition walls subdividing the interior of a vessel, closure gate, or floating dry dock into separate compartments, or (3) a retaining wall used as a form of coastal management, akin to a seawall.

Buoyancy: The upward push of water pressure, equal to the weight of the volume of water that the vessel or dry dock displaces.

Capstan: Steel warping drum that rotates on a vertical axis for the handling of mooring lines.

Caisson (Caisson Gate): (1) The floating entrance gate to a graving (basin) dry dock, or (2) a cast in place concrete pile/column, usually of large diameter.

Cathodic Protection: A method to reduce corrosion of steel structures by minimizing the difference in potential between the anode and cathode. The two main types used in dry docks are *sacrificial anode protection systems*, which utilize sacrificial aluminum, magnesium, or zinc anodes; and *impressed current cathodic protection systems* (ICCP), which utilize current flow to the anode from an external power source.

Center of Buoyancy: The centroid of the submerged volume of a vessel or vessel/dry dock system.

Center of Gravity: The location at which an object's weight could be replaced by a concentrated force without changing its behavior.

Certified Rated Capacity: The maximum certified load which the dry dock can structurally lift. For floating dry docks, this is the vessel displacement weight that the dry dock can lift and support in a satisfactory condition at the minimum required pontoon freeboard.

Certified Rated Keel Load: The maximum certified vessel keel load that can be applied to the dry dock.

Certifying Agency: An engineer or naval architect (individual or firm) registered or licensed to practice professional engineering or naval architecture, as defined by the statutory requirements of the professional registration laws of the state in which the drydocking facility is located.

Cleat: A low-capacity mooring fitting with two diverging horizontal horns around which dry dock lines are wrapped.

Cradle: A structure used to support a vessel.

Curves of Form: *See* **Hydrostatic Curves**.

Deflection Monitoring System: (1) Typically, an optical, sight, or laser, arrangement of targets used to determine the bending of a floating dry dock, or (2) draft or stress monitoring system.

Displacement: All-inclusive weight of a vessel or floating structure equal to the weight of the water displaced.

Dockmaster: Individual responsible for the safe operation of the dry dock and the safety of the vessel during drydocking operations and the lay period.

Draft: The vertical distance from the keel to the waterline.

Dry Dock: (Noun) General term for a drydocking facility used for the inspection, maintenance, repair, and/or construction of vessels.

Dry Dock Draft (Floating Dry Docks): The vertical distance from the lowest point on a dock, typically the keel, to the waterline.

Dry Dock Translation System: A mechanism used to change the location of a floating dry dock.

Drydocking (Drydock, Drydocked): (Verb) The action of bringing a vessel into a dry dock and landing the ship on the supporting blocks/structures.

Drydocking Evolution: The act of bringing a vessel into a dry dock from preparation through the final condition in which work will be performed.

Drydocking Facility: (Noun) General term for a floating dry dock, graving dock, marine railway, or vertical lift.

Drydocking Plan: Detailed structural plan and profile of the lower hull structure required for correct support of the vessel in drydocking.

Fender: A device placed between a dock structure and a vessel, used to absorb berthing impact energy and provide a wearing surface for the vessel while moored to the structure.

Fendering Timber: An intermediate member that is used to transfer load between a vessel and a dry dock or pier to prevent damage to the vessel or structure. The material is typically wood, plastic, or rubber.

Freeboard: The elevation above the waterline of a floating dry dock's pontoon deck.

GM: *See* **Metacentric Height**.

Hauling Side Blocks: Supports used to prevent the overturning of a vessel in dry dock. They can be positioned against the hull during the drydocking evolution.

Hydrographic Survey: An assessment of a water-covered area to determine the depth of water relative to a reference plane (i.e., mean low water).

Hydrostatic Curves: Also referred to as *curves of form* or *displacement and other curves* (D&O curves). A graphical description of properties of the vessel's form, which are plotted against the draft of the vessel. Properties that are always included in the hydrostatic curves are displacement, longitudinal center of buoyancy (LCB), height of transverse metacenter above keel (KM), tons per inch (centimeter) immersion (TPI) (TPcm), longitudinal center of flotation (LCF), and moment to trim 1 in. (2.54 cm) (MTI) (MTcm).

Hydrostatic Forces: *See* **Hydrostatic Pressure**.

Hydrostatic Head: A pressure caused by the height of the liquid above the point of interest.

Hydrostatic Pressure: The pressure at a given depth in a liquid caused from the weight of the liquid acting on a unit area at that depth.

ICCP: Impressed Current Cathodic Protection System. *See* **Cathodic Protection**.

KB: The vertical distance from the keel of a ship to the center of buoyancy.

KG: The vertical distance from the keel of a ship to the ship's center of gravity.

Keel: The principal bottom structural element of a ship extending along the centerline along the length of the vessel.

Keel Block(s): Movable structural supports that are aligned under the keel of a vessel to be drydocked and above the dry dock structure to transfer the deadweight of a vessel to the dry dock structure.

KM: The vertical distance from the keel of a ship to its metacenter. The metacenter is considered to be fixed for small angles of heel.

Knuckle Block (Suing Block): The first keel block on which the vessel knuckles (sues).

Knuckle Reaction (Suing Reaction): The force that develops at the knuckle block, on which the ship rotates, until it makes contact with all the keel blocks.

Landed: The point at which the keel has made contact with all the keel blocks.

Lift Capacity (Maximum): The vessel displacement weight that the dry dock can lift and support in a satisfactory condition. For floating dry docks, this must include the minimum pontoon freeboard.

Light Dock Survey: The process of recording a floating dry dock's floating condition and the mass, longitudinal, transverse, and vertical location of each item which should be added, deducted, or relocated on the floating dry dock based on its current condition. The light dock survey is also performed to calculate the **light dock weight** of the floating dry dock.

Light Dock Weight: The normal, unloaded state of a floating dry dock with all operating equipment on board and the water level in all tanks as low as the dewatering pumps can obtain.

Light Ship Weight: The weight of the complete ship including hull, machinery, outfit, equipment, and liquids in machinery, but not including any variable or movable weights such as ballast water, cargo, fuel, or stores.

List: A measure of the deviation from vertical of a ship's vertical axis in the athwartship direction. Measured as a difference in the port and starboard draft readings.

Longitudinal Center of Buoyancy (LCB): The longitudinal center of the displaced volume referenced to either the forward reference point or amidships.

Longitudinal Center of Flotation (LCF): The longitudinal centroid of the waterplane, referenced to either the forward reference point or amidships. It is the point about which the vessel trims as a weight is shifted longitudinally aboard the vessel.

Longitudinal Strength: The maximum capacity of a structure to resist global shear and bending moment owing to uneven distribution of loads, such as buoyancy and weights, along its length.

Margin Line: A line 3 in. (76 mm) below the top wingwall deck or the lowest nonwatertight wingwall penetration of a floating dry dock, defining the highest permissible location of any water plane in the final condition of sinkage, trim, and heel following damage.

Metacenter: The point through which a vertical line from a ship's center of buoyancy as the ship rolls intersects the original vertical centerline of the ship.

Metacentric Height (GM): The vertical distance from a vessel's center of gravity (G) to its metacenter (M). Provides measure of the initial stability of a vessel.

MHW: (Mean High Water) The mean of all high water levels (two per day for diurnal tides), observed over a 19-year tidal epoch (Metonic cycle).

MHHW: (Mean Higher High Water) The mean of the highest daily highwater levels (one per day), observed over a 19-year tidal epoch (Metonic cycle).

MLW: (Mean Low Water) The mean of all low water levels (two per day for diurnal tides), observed over a 19-year tidal epoch (Metonic cycle).

MLLW: (Mean Lower Low Water) The mean of lowest low water levels (one per day), over a 19-year tidal epoch (Metonic cycle).

Moment Capacity: The maximum flexural strength that a structural member can safely support under load.

Moment Connections: An attachment between two structural members that is capable of transferring a moment.

Moment to Trim 1 In. (MT1") (MTcm-metric): A hydrostatic property of the vessel indicating the necessary differential variation of weight acting at a distance required to influence the trim of the vessel by 1 in. (2.54 cm).

Mooring Dolphin: A structure used to secure a vessel on location. This typically comprises a series of driven piles.

MSL: (Mean Sea Level) The average height of the sea surface for all stages of the tide, measured hourly, over a 19-year tidal epoch (Metonic cycle).

MTL: (Midtide Level) The elevation midway between mean high water and mean low water. Also called half-tide level, it may not be the same as mean sea level.

Naval Architect: An individual trained in the design, construction, and repair of marine vessels.

Overturning Moment: The result of a load located or perturbed away from the center of gravity that tends to increase the list of a vessel.

Pile: An in-ground column, typically of timber, reinforced plastic, steel, or concrete, driven or otherwise set into the soil or rock to resist applied vertical and/or horizontal loads.

Pontoon: The structure that extends between and under the wingwalls to form the bottom of the floating dry dock.

Rated Lifting Capacity: The ship weight that the dry dock can lift and support in a satisfactory condition (at the rated pontoon freeboard for floating dry docks).

Righting Moment: A vessel's ability to right itself, resulting from an offset in location of the center of buoyancy and center of gravity.

Sacrificial Anode: An anode that provides its own electrical current for cathodic protection of a steel structure by consuming or "sacrificing" itself. These anodes may be formed from various metals and alloys, but often mostly from zinc or aluminum.

Safety Deck: (1) An intermediate watertight deck within the wingwall of a floating dry dock, forming the upper boundary of the ballast tanks. In a properly designed floating dry dock, the elevation of the safety deck ensures sufficient buoyancy to keep the dock afloat when all ballast tanks are flooded, or (2) graving dock caisson gates are also designed with safety decks.

Sea Chest: An enclosure, attached to the inside of the underwater shell and open to the sea, fitted with a strainer plate. A sea valve and piping connected to the sea chest passes seawater into the vessel for ballasting, cooling, or firefighting purposes.

Ship Owner Representative: Company or designated representative that is responsible for the vessel.

Shipyard: (1) Location that contains a dry dock where vessels are repaired and/or built, or (2) the owner or tenant responsible for the maintenance and operation of a drydocking facility.

Shipyard Management: Management team responsible for the safe operation of the drydocking facility.

Shore/Shoring: A structural member, such as a strut, used to brace a vessel in a dry dock.

Side Blocks: Movable structural supports that are aligned away from the keel of a vessel to be drydocked and above the dry dock structure to transfer the deadweight of a vessel to the dry dock structure.

Splash Zone: The portion of a marine structure subject to intermittent wetting and high chloride exposure (seawater) owing to wave action and spray. Typically, a zone of high steel corrosion levels.

Stern Reference Point (SRP): A plane at the stern of the ship used as a reference for longitudinal measurement. Usually, the extreme aft portion of the vessel and identified on the docking plan.

Superflood: The act of pumping water into the dry dock, such that the water level inside the basin is higher than the adjacent waterway.

Tank Level Indicator (TLI): A means of measuring the water level within a tank.

Tons Per Inch Immersion (TPI) (TPcm-metric): A hydrostatic property of the vessel indicating the differential change in displacement for every inch (2.54 cm) the vessel is submerged.

Transfer: The physical act of moving a vessel from landside to a dry dock or vice versa.

Translate (Translation): The physical act of moving a floating dry dock from one position to a second position. For example, moving the floating dry dock from its moored position to a vessel transfer position at a bulkhead.

Transverse Bending Strength: The capacity of a structure to resist a transverse bending moment. *See* **Moment Capacity**.

Transverse Center of Gravity (TCG): The athwartship (port or starboard) location of the center of gravity measured from the longitudinal center line of a vessel.

Transverse Inertia: The moment of inertia of a structural member that would be effective for a transverse bending moment.

Transverse Strength: The maximum capacity of a structure to resist loads that are unevenly distributed across its width, such as block loads or the concentrated weight of a floating dry dock's wingwalls.

Trim: The difference in the forward and aft draft of a ship.

Trim and Stability Booklet (T&S Book): A handbook placed aboard a vessel for use by the vessel operators to assist in calculating the stability of the vessel. The document consists of tables and drawings indicating the essential characteristics of all spaces aboard a vessel where loads may be placed.

Ultrasonic Thickness (UT): A nondestructive technique used to measure (gauge) the thickness of a steel member using sound waves.

Undocking: (Verb) The action floating a vessel off its support blocks/structures and moving the vessel out of a dry dock.

Vertical Center of Gravity (VCG or KG): The vertical location of the center of gravity measured as a distance above the keel.

Weather Deck: Uppermost hull, closure gate, or floating dry dock deck exposed to the weather at all times.

Weep Hole: Opening provided in a wall or bulkhead to facilitate drainage of water. It usually serves to reduce hydrostatic pressure behind the structure.

Wildcat: A pocketed and slotted wheel on a winch over which a chain passes.

Winch (or Windlass): An engine fitted with a rotating drum for hauling ropes. Some are fitted with multiple drums, a gypsy head for hauling ropes, or a wildcat for hauling chains.

Wing Tank: Ballast tank adjacent to the hull side.

1.4 CONSENSUS STANDARDS AND OTHER REFERENCED DOCUMENTS

This section lists the consensus standards and other documents that are adopted by reference within this standard.

OSHA (Occupation Safety and Health Administration)

OSHA 2268-10R2014, *Shipyard Industry Standards*.

CHAPTER 2
TYPES OF DRY DOCKS

2.1 PURPOSE

This section provides basic information about the different types of dry docks covered by this standard.

2.2 FLOATING DRY DOCKS

Floating dry docks, as shown in Figures 2-1 and 2-2, are structures with sufficient size, strength, displacement, and stability to lift a vessel from the water using buoyancy. Floating dry docks can be operated with list and trim to reduce block loading and/or to reduce or eliminate vessel stability problems when drydocking or undocking. Some shipyards are equipped to transfer vessels between the floating dry dock and a shore facility, which enables concurrent work on multiple vessels.

Floating dry docks are comprised of a pontoon and wingwalls. The *pontoon* is the main structural component that distributes the concentrated blocking loads from the vessel to the dry dock. The pontoon must have sufficient volume to provide the buoyancy to lift the vessel and the floating dry dock out of the water. The *wingwalls* provide stability when the pontoon is submerged, and longitudinal strength to distribute the vessel's weight to the uniform buoyant support provided by the pontoon. Floating dry docks can be subdivided into the following three categories based on their structural arrangement: pontoon, caisson, or sectional floating dry docks.

2.2.1 Pontoon Floating Dry Dock A *pontoon floating dry dock*, also referred to as a *Rennie type* floating dry dock, is comprised of continuous wingwalls spanning a series of detachable pontoons. The discontinuous pontoon deck results in reduced longitudinal stiffness. Typically, the pontoon sections are designed to be self-docking for inspection and maintenance.

2.2.2 Caisson Floating Dry Dock A *caisson floating dry dock*, also referred to as a *box* or *one-piece floating* dry dock, is

Figure 2-1. Floating dry dock.
Source: US Navy photo by Mass Communications Specialist Seaman Kelly E. Barnes.

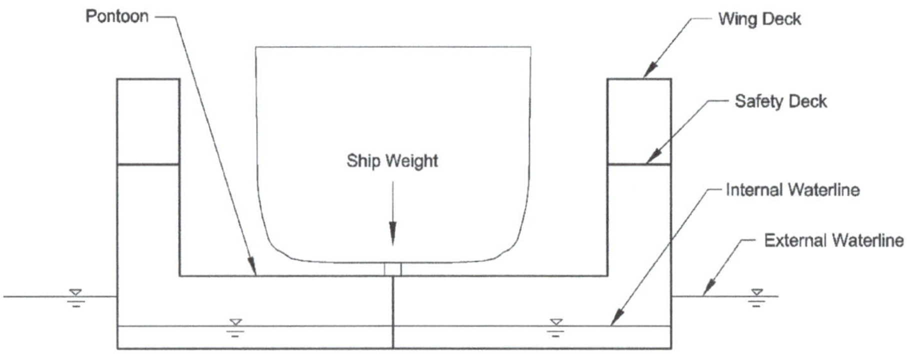

Figure 2-2. Floating dry dock components.

comprised of continuous wingwalls connected to the pontoon. This type of dry dock is typically lighter and stronger than other types of floating dry docks because the full lengths of both the pontoon and wingwalls are effective for longitudinal strength. Caisson floating dry docks are not self-docking and must be drydocked in a larger drydocking facility for inspection and maintenance.

2.2.3 Sectional Floating Dry Dock A *sectional floating dry dock* is comprised of a series of sections, each containing a pontoon and wingwalls. The sections are joined with locking lugs, hinge pins, or moment connections. Locking logs loosely align the sections but typically cannot support significant load. Hinge pins provide shear strength but have no moment capacity. Sectional floating dry docks with moment connections at the top and bottom of the wingwalls act like pontoon floating dry docks. Typically, sectional floating dry docks are designed to be self-docking for inspection and maintenance.

2.3 GRAVING DOCKS

Graving docks, also referred to as *basin dry docks*, are dry docks with a basin floor located below the adjacent water level. The basic structure of a graving dock, such as the one shown in Figure 2-3, consists of a floor, basin sidewalls, a headwall, and a closure gate. Graving docks can be further subdivided into the following three categories, based on the means used to resist the buoyant uplift force: full hydrostatic, fully relieved, or partially relieved graving docks.

Figure 2-3. Graving dock.
Source: Courtesy of General Dynamics Electric Boat, Groton, Connecticut.

2.3.1 Full Hydrostatic Graving Dock A *full hydrostatic graving dock* relies either on its own weight or on an anchorage system to resist the buoyant uplift forces acting on the graving dock. Regardless of the surrounding geology, it is conservative to assume that the graving dock structure will be exposed to full hydrostatic pressure. For this type of graving dock, the full buoyant force at the highest water table elevation must be resisted by either the weight of the structure, the combined weight of the structure and foundation material (rock, soil, etc.) engaged by hold-down devices such as anchors or piles, the weight of soil resting on the ledge formed by projections of the floor beyond the basin sidewalls, the friction created by the soil on the basin sidewalls, or a combination of these. A schematic section view of a full hydrostatic graving dock is shown in Figure 2-4.

2.3.2 Fully Relieved Graving Dock A *fully relieved graving dock* does not have sufficient weight to resist the buoyant uplift forces acting on the graving dock. This type of graving dock does not require as much material to construct, because it relies on a drainage system to remove the surrounding water behind the basin sidewalls and beneath the basin floor to alleviate the hydrostatic pressure. However, it does require the continuous operation and maintenance of the drainage system throughout the life of the dry dock. A schematic section view of a fully relieved graving dock is shown in Figure 2-5.

2.3.3 Partially Relieved Graving Dock A *partially relieved graving dock* requires relief of the hydrostatic force either from under the floor slab, or from behind the basin sidewalls. For graving docks where only the floor is relieved, such as the graving dock shown in Figure 2-6, the basin sidewalls are designed to resist the full hydrostatic head pressure. In this scenario, a sheet pile cutoff wall is typically driven at the perimeter of the floor slab to reduce seepage under the slab. For graving docks where only the basin sidewalls are relieved, the floor slab is designed to resist the full buoyant uplift force. Partially relieved graving docks require the continuous operation and maintenance of the drainage system throughout the life of the dry dock.

2.3.4 Closure Gates All graving docks have a *closure gate* located at the entrance of the basin. When the closure gate is placed in its seated (closed) position, it creates a watertight seal, which allows the dry dock basin to be dewatered or superflooded. During drydocking operations, the closure gate is removed or rotated away from its seat to allow a vessel to enter or exit the

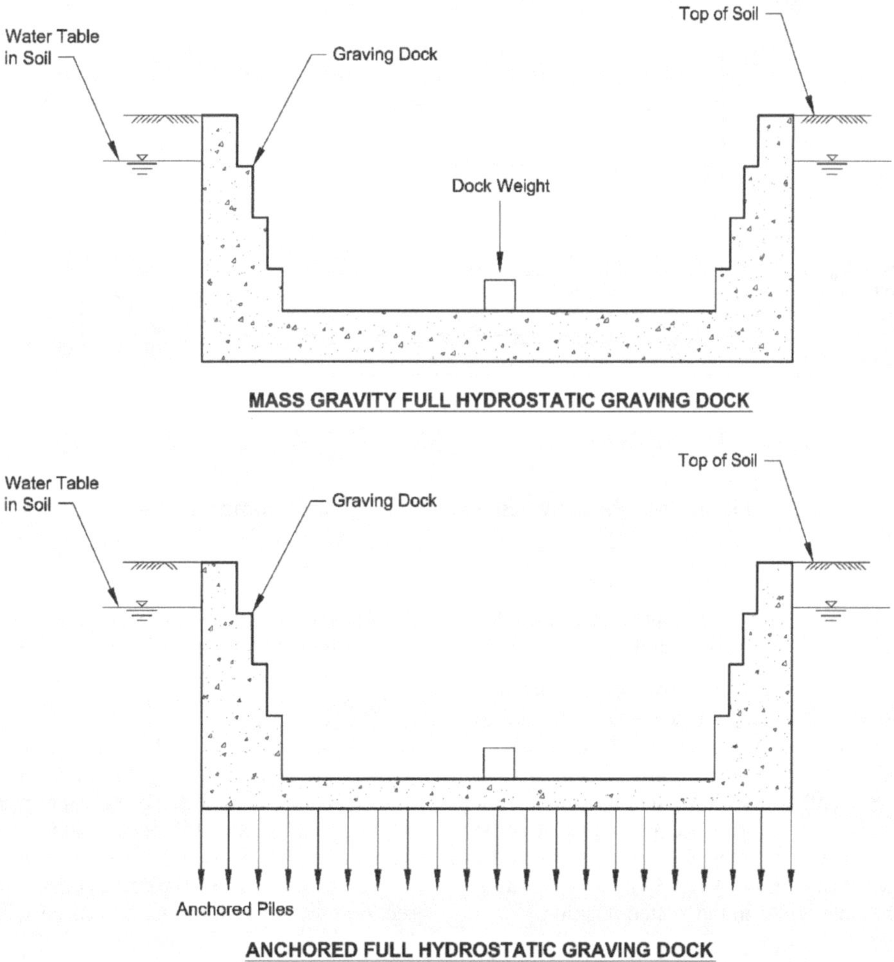

Figure 2-4. Full hydrostatic graving dock components.

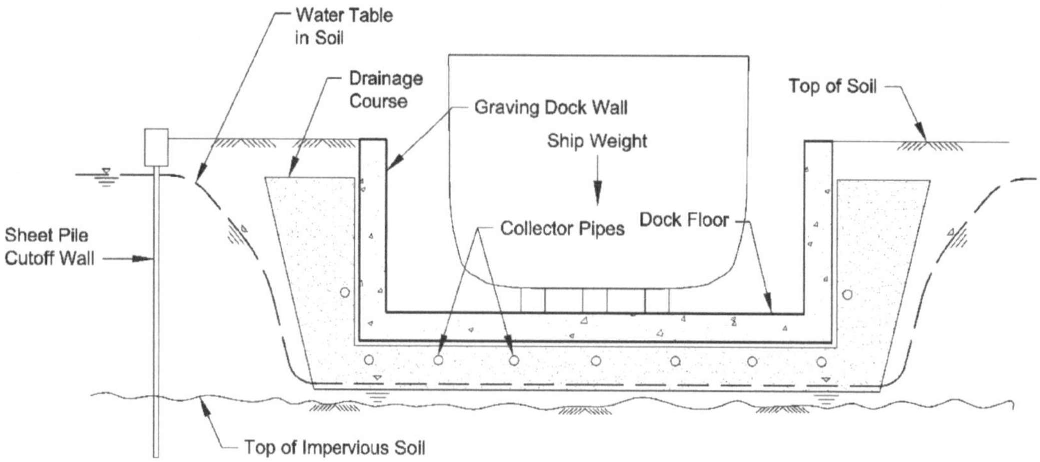

Figure 2-5. Fully relieved graving dock components.

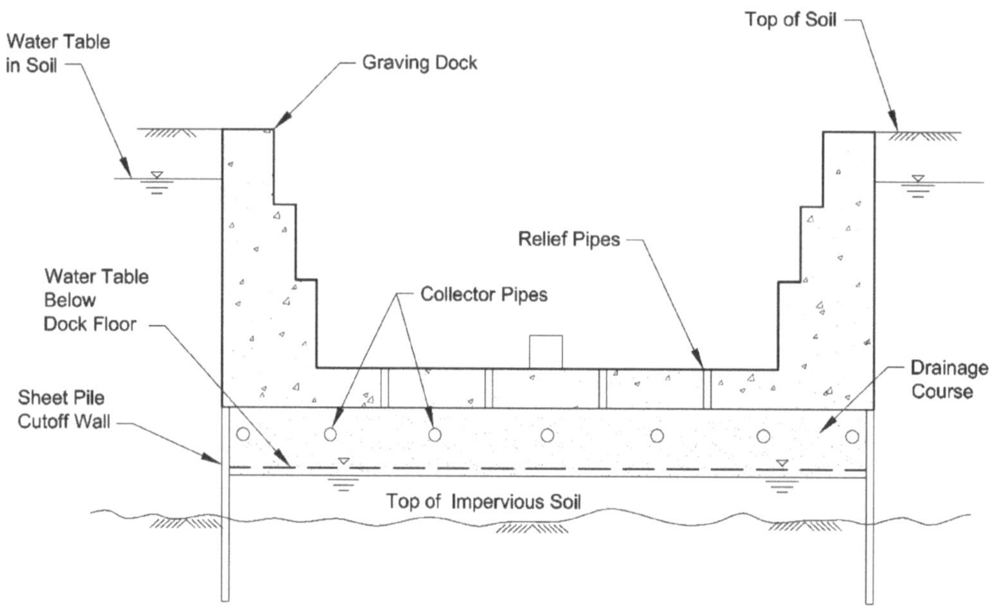

Figure 2-6. Partially relieved graving dock components.

graving dock. There are two types of closure gates used for graving docks—caisson gates and hinged gates.

2.3.4.1 Caisson Gate A *caisson gate* is a structure that is either floated away from its seat or rolled into a notch in the basin sidewall. The floating caisson gate is the most widely used closure gate for graving docks and is constructed with an internal watertight tank or tanks, outfitted with valves and pumps for ballasting and deballasting operations. A caisson gate may be made to be reversible, such that either side of a gate may be presented to the dry dock basin, allowing flexibility for inspection and maintenance of the gate while in the seated position.

2.3.4.2 Hinged Gate A *hinged gate* is a structure that is rotated away from the seat with either side or bottom hinges. Opening or closing of the hinged gate is typically achieved and controlled using ropes or wires driven by a winch.

2.3.4.3 Intermediate Gates *Intermediate gates* are structures located inside the graving dock between the closure gate and the headwall. Placement of the intermediate gate allows the inboard end of the dry dock basin to remain dry, while the outboard end is flooded. Intermediate gates utilize a seat similar to a closure gate, and typically consist of multiple overlapping sections, which can be put into place with a crane.

2.4 MARINE RAILWAYS

A *marine railway* is a mechanical means of hoisting a ship out of the water along an inclined track. This type of dry dock, as shown in Figures 2-7 and 2-8, consists of a drydocking cradle that rides on inclined rails that extend from the shore into the water. The cradle is lowered down the track by a backing chain (or wire rope) assisted by gravity. The cradle is moved up the track by a hauling chain (or wire rope) driven by a winch and a series of gears.

2.5 VERTICAL LIFTS

A *vertical lift* is a mechanical device for hoisting a vessel out of the water vertically. It consists of a horizontal lifting platform suspended from a series of synchronous hoists, arranged in pairs along both sides of the platform, which are supported on fixed piers, as shown in Figures 2-9 and 2-10. The platform is fabricated from transverse and longitudinal steel beams. During drydocking operations, keel blocks and side blocks are typically secured directly to the platform or onto rail mounted, wheeled

Figure 2-7. Marine railway.
Source: US Coast Guard, photo by PA2 Kyle Niemi.

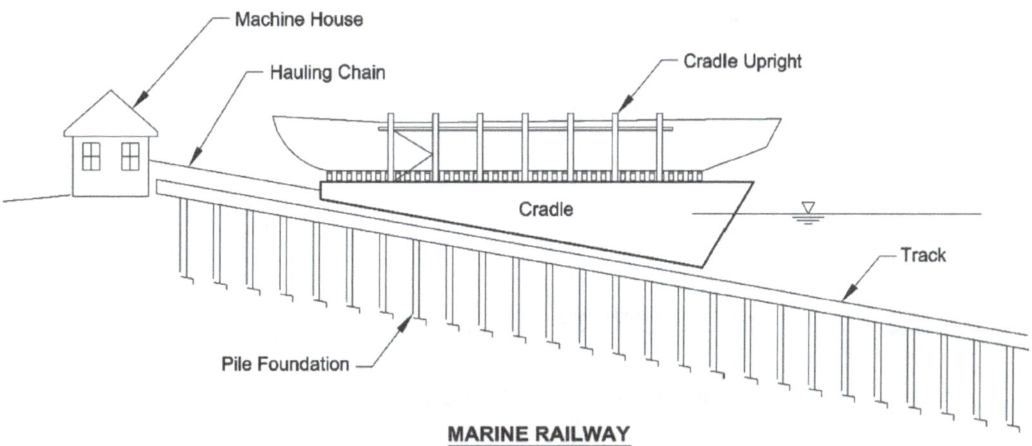

Figure 2-8. Marine railway components.

Figure 2-9. Vertical lift.
Source: Courtesy of Heger Dry Dock, Holliston, Massachusetts.

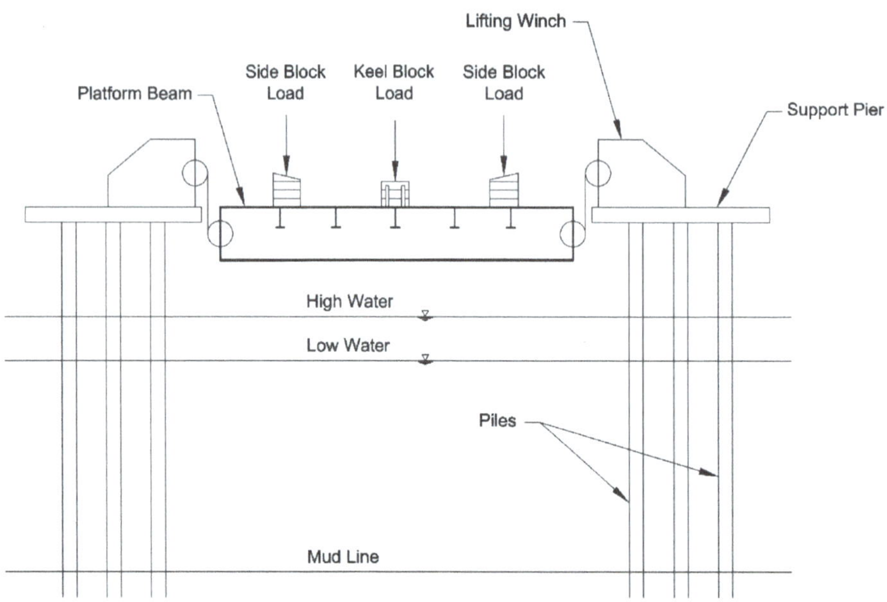

Figure 2-10. Vertical lift components.

cradles. The synchronized hoists lower the platform into the water until there is sufficient draft over the blocks, and the vessel is positioned over the blocks and secured to the hoist piers with conventional mooring lines. The platform is then raised to shipyard grade. Most shipyards are equipped to transfer vessels off the lifting platform to a land-level shore berth on the wheeled cradles, which enables concurrent work on multiple vessels. For undocking operations, the synchronized hoists lower the platform, cradle, and vessel into the water, and the vessel is then floated off the platform.

CHAPTER 3
DRY DOCK CERTIFICATION REQUIREMENTS

3.1 REQUIREMENTS

To obtain and sustain dry dock certification in accordance with this standard the shipyard shall

1. Maintain documents governing the drydocking facility's design, capacity, stability (as applicable), geometry, construction, modification, maintenance, repair, operation, and inspection reports. These documents shall be made available to the certifying agency for their use during the certification process.
2. Establish and follow a maintenance program.
3. Operate the dry dock in accordance with this standard.
4. Assign a qualified Dockmaster for drydocking evolutions.
5. Maintain a log in accordance with this standard of all drydocked vessels.
6. Have a certifying agency conduct a condition assessment to establish, verify, or adjust the dry dock's certified rated capacity and stability limitations (as applicable). The certifying agency shall establish, verify, or adjust the safe operational limitations of the dry dock.
7. Have a certifying agency conduct periodic certification inspections.
8. Provide full access to all areas of the drydocking facility to allow for 100% visual inspection by the certifying agency.
9. Operate the dry dock in the presence of the certifying agency.

On completion of the condition assessment, dry dock certification in the form of a Dry Dock Capacity Certificate shall be issued for the dry dock's certified rated capacity and stability limitations (as applicable) by the certifying agency who conducted the condition assessment.

3.2 QUALIFICATIONS OF THE DRY DOCK CERTIFYING AGENCY

Dry dock condition assessments and subsequent certification inspections shall be conducted by a dry dock certifying agency. The certifying agency is defined as an engineer or naval architect (individual or firm) licensed to practice professional engineering, as defined by the statutory requirements of the professional registration laws of the state in which the drydocking facility is located. The certifying agency shall be individually qualified for its respective role in dry dock material condition surveys and evaluations. The certifying agency shall not be personnel who are undertaking routine maintenance or operation of the dry dock.

All certifying agency personnel involved in the condition assessment or certification inspection shall possess the technical qualifications, including relevant experience, education, and training, required to perform the individual tasks assigned. An individual team member for the underwater portion of the condition assessment or certification inspection shall have experience in one or more of the following areas: underwater inspection, underwater salvage work, underwater ship surveying, or underwater maintenance and repair of materials appropriate to the facility. A diver meeting the requirements of a recognized classification society for similar underwater inspection shall be considered to be qualified in accordance with the aforementioned requirements to perform the underwater portion of the inspection.

3.3 DRY DOCK CAPACITY CERTIFICATE

The Dry Dock Capacity Certificate shall be signed and sealed by a licensed professional engineer experienced with the type of drydocking facility being certified. The Dry Dock Capacity Certificate shall include the following minimum information:

1. Certified rated keel load (load per unit length),
2. Certified rated capacity of the dry dock,
3. List of any items on which certification is contingent with justification/reason(s),
4. Date of certification,
5. Expiration date of certification,
6. Name of the certifying agency,
7. Signature of the certifying agent(s), and
8. List of any operational restrictions with justification/reason(s).

For floating dry docks and marine railways, the Dry Dock Capacity Certificate shall include reference to the applicable dry dock system (vessel and dry dock) stability calculations and limitations.

The Dry Dock Capacity Certificate shall be posted at the central operations station of the drydocking facility.

3.4 CONDITION ASSESSMENT

A dry dock condition assessment shall be conducted by a certifying agency for each dry dock that intends to obtain dry dock certification status. The certifying agency shall gather all pertinent design data on the drydocking facility, review the design data, conduct a material condition survey to establish the present (existing) baseline condition of the dry dock, witness operation of the dry dock, and review or perform calculations (if necessary) to verify (or establish) the certified rated capacity, the stability limitations (as applicable), and the operational limitations of the dry docking facility. The certified rated capacity, the stability limitations, and the operational limitations shall be used by the Dockmaster and/or the Dry Dock Engineer to ensure that the dry dock's structural capacity, buoyant capacity (as applicable), stability limitations (as applicable), and operational capabilities are not exceeded.

The scope of work required for the condition assessment will vary, depending on the type of facility being evaluated. Detailed condition assessment requirements for each dry dock type are outlined in the following chapters of this standard:

Floating Dry Docks, Chapter 4,
Graving Docks, Chapter 5,
Marine Railways, Chapter 6, and
Vertical Lifts, Chapter 7.

3.4.1 Periodicity of Condition Assessments A dry dock condition assessment shall be performed when

1. Dry dock is not currently certified in accordance with this standard.
2. Dry dock is new.
3. Dry dock has been relocated (as applicable).
4. Dry dock condition assessment shall also be performed when an action or event affects the dry dock's capacity, stability (as applicable), operational limitations, or operation, including but not limited, to the following:
   (a) Dry dock facility has been modified.
   (b) Shipyard maintenance activities discover deficiencies or damage.
   (c) Certification Inspections discover deficiencies.
   (d) Dry dock facility has been damaged such that its operations or capacity have been affected.

3.4.2 Drawings Where drawings exist, the certifying agency shall verify their accuracy. If the drawings are found to be accurate and complete, the drawings shall be used for the condition assessment.

Where design drawings are not available or incomplete, design characteristics shall be determined and documented. At a minimum, the following shall be documented:

- General arrangement;
- Structural characteristics;
- Available landside utilities required to operate the dry dock; and
- Other characteristics required to determine the capacity, stability (as applicable), operational limitations, and operation of the dry dock.

3.4.3 Dry Dock Calculations Where structural capacity calculations, buoyancy calculations (as applicable), stability calculations (as applicable), and operational limitations from the original design exist, the certifying agency shall verify their accuracy. If the calculations are found to be accurate and complete, the calculations shall be used for the condition assessment.

Where structural capacity calculations, buoyancy calculations (as applicable), stability calculations (as applicable), and operational limitations from the original design are not available or incomplete, the certifying agency shall establish the structural capacity, buoyant capacity (as applicable), and stability limitations (as applicable) of the dry dock.

Calculations that support the operational limitations of the dry dock shall be based on the existing conditions of the drydocking facility. If the dry dock has deteriorated to the point where the structural capacity, buoyant capacity (as applicable), and/or stability (as applicable) are in question, then new dry dock calculations shall be performed and new operational limitations established.

3.4.4 Material Condition Survey A material condition survey is required to establish the existing condition of the drydocking facility. The certifying agency shall ascertain the existing condition of all critical components of the facility. All findings and deficiencies shall be documented on the certification inspection checklists and included in the Condition Assessment Report. Representative examples of certification inspection checklists are provided in Appendix A for each dry dock type.

In areas where section loss is suspected, such as areas of failed or missing protective coating, areas of corrosion, or where the as-built member thicknesses are in question, ultrasonic thickness (UT) measurements, calipers, or other devices shall be used to determine the remaining structure thickness. The certifying agency responsible for the condition assessment shall establish the frequency and location of the measurements, which will vary depending on the condition of the structure.

3.4.5 Condition Assessment Report On completion of the condition assessment, the certifying agency shall prepare a written report summarizing the findings of the condition assessment. At a minimum, the Condition Assessment Report shall include the following:

1. Scope of the condition assessment performed;
2. Experience and qualifications of the certifying agency and key personnel who performed the condition assessment;
3. Pertinent history that is available on the dry dock; this shall include name of designer, name of builder, date of construction, major dry dock overhauls, or renovations;
4. List of major storms, hurricanes, earthquakes, or accidents that caused damage to, and the extent of the damage to, the dry dock.
5. List of dry dock relocations (if applicable);
6. Number of hours or operational cycles as appropriate;
7. List of vessels drydocked;
8. Written description of the following:
   (a) Dry dock's composition,
   (b) Water levels at the dry dock with reference to a datum,
   (c) Vessel support blocking system to which the dry dock is certified,
   (d) Dry dock's operation,
   (e) Closure gate's operation (if applicable),
   (f) Ship handling system,
   (g) Electrical power system,
   (h) Dry dock's control and monitoring systems, and
   (i) Fire protection system.
9. Summary of the material condition survey:
   (a) Include all findings in the certification inspection checklists.
   (b) Include dates of the material condition survey.
   (c) Provide a list of items not inspected and reasons for lack of inspection.
   (d) Provide a list of equipment which was not operated and reasons for lack of operation.
   (e) Provide a list of all MARGINAL conditions along with a disposition.
   (f) Provide a list of all UNSATISFACTORY conditions along with a disposition.
   (g) Include pertinent photographs and/or video recordings.
10. List of documents and calculations reviewed and/or developed to verify the dry dock's capacity, stability (as applicable), and operational limitations.

11. Summary of the stability limitations (as applicable) of the dry dock system (vessel and dry dock) based on the parameters in this standard.
12. Summary of the operational limitations of the dry dock.
13. Dry Dock Capacity Certificate.
14. Recommendation regarding the frequency and scope of future certification inspection surveys based on the condition of the dry dock and the operating environment.

3.5 CERTIFICATION INSPECTIONS

A dry dock certification inspection shall be conducted by a certifying agency for each dry dock to sustain dry dock certification status. The certification inspection shall be conducted prior to the expiration of the previous certification period to maintain active certification status. Detailed certification inspection requirements are outlined in Chapter 8 of this standard.

3.6 MATERIAL CONDITION SURVEY REPORTING

3.6.1 Reporting Certification inspection checklists shall be used to record the findings of the material condition survey. The checklists shall be modified to suit each facility and shall have sufficient detail to ensure representative inspection of all structures, as well as mechanical and electrical equipment. Structural members such as transverse and longitudinal strength members, and plating and framing for decks, bulkheads, and shells shall be listed individually by ballast tank, compartment, frame number, or other identifier. Mechanical and electrical equipment such as valves, pumps, and sluice gates shall also be listed individually. The grouping of items for an entire facility, such as "pontoon structure," "marine railway cradle structure," or "ballast valves," is insufficient. The necessity or urgency for correcting deficiencies shall be explained on the checklists. The following material condition ratings shall be utilized:

1. SATISFACTORY (SAT or S): The condition of the item will not result in system damage and based on measured or estimated deterioration rate; it may be expected to remain SATISFACTORY until the next certification inspection.
2. MARGINAL (M): The condition of the item will not result in major damage or, by itself, will not make the drydocking facility unsafe to drydock a vessel within the dry dock's capacity, provided it is corrected, repaired, or replaced in a timely manner defined by the dry dock certifying agency. A number of such items as a group can make the drydocking facility unsafe. This shall be evaluated by the surveyor.
3. UNSATISFACTORY (UNSAT or U): The condition of the item may cause system damage or loss. The item shall be corrected, repaired, or replaced immediately if there is a vessel in the dry dock and the work can be done safely. If there is no vessel in the dry dock then the item shall be corrected, repaired, or replaced prior to drydocking a vessel.
4. NOT APPLICABLE (NA): The item listed in the checklist is not present on the dry dock. Note that as items are added, removed, or replaced from the dry dock, the certification inspection checklist shall be updated to reflect the configuration changes.
5. NOT INSPECTED (NI): The item was not inspected as part of the current certification inspection interval. It is not acceptable to mark an item NI unless there is a basis for expecting the item to be in a SATISFACTORY condition. Items shall not be marked NI until the certification inspection for that interval has been completed. Items shall not be marked NI solely for convenience. Any items that were inaccessible during the current certification inspection shall be prioritized and documented based on their level of risk.
6. Items marked MARGINAL, UNSATISFACTORY, NOT APPLICABLE, or NOT INSPECTED shall include notes or other documentation clearly stating the deficiency and the reasons for less than SATISFACTORY condition rating.

3.7 SPECIFIC REQUIREMENTS FOR ALL DRY DOCK TYPES

3.7.1 Dry Dock Maintenance The shipyard shall develop and follow a dry dock maintenance program. The maintenance program shall schedule dry dock maintenance, track repairs, and maintain records. Dry dock maintenance requirements are included in Chapter 9 of this standard.

3.7.2 Dry Dock Operations The shipyard shall assign a Dockmaster for each drydocking evolution and shall operate the dry dock in accordance with this standard. Detailed dry dock operation requirements, including qualifications requirements of the Dockmaster and dry dock operations personnel, are outlined in Chapters 10, 11, and 12 of this standard.

3.7.3 Fire Protection The shipyard shall have an installed fire protection system compliant with Occupational Safety and Health Administration (OSHA) regulations. If a fire protection system is not installed, the shipyard shall secure and maintain for records a *Memorandum of Agreement* with a local fire department ensuring that the fire department can arrive at the facility within 15 min of receiving the alarm. The *Memorandum of Agreement* shall be dated within 5 years of the date of the Dry Dock Capacity Certificate.

3.7.4 Hydrographic Survey A hydrographic survey shall be conducted every 5 years unless local site conditions justify a longer or shorter frequency. The hydrographic survey shall show available water depths at the drydocking facility, and at the approach channel. The survey datum shall be clearly labeled with reference to water levels and to shipyard grade.

3.7.5 Underwater Survey An underwater survey shall be conducted every 5 years.

3.7.6 Record Keeping The shipyard shall keep and maintain records as outlined in Chapter 13 of this standard.

3.8 SPECIFIC REQUIREMENTS FOR FLOATING DRY DOCKS

3.8.1 Instrumentation Requirements Floating dry docks must be outfitted with the following:

1. Ballast tanks must be outfitted with a water level sensing system.
2. Wingwalls must be outfitted with draft boards at each corner. For dry docks longer than 300 ft (91 m), draft boards must also be located at the longitudinal center of each wingwall.
3. Longitudinal deflection monitoring system for docks 300 ft (91 m) and longer.

3.8.2 Wingwall Freeboard When the floating dry dock is submerged to its maximum draft, the wingwall freeboard shall be a minimum of 3 ft (0.914 m), measured from the top of the wingwall weather deck to the wingwall margin line.

3.8.3 Pontoon Freeboard The floating dry dock pontoon deck freeboard at the operating draft (operating pontoon freeboard) shall be a minimum of 12 in. (305 mm), measured from the lowest point of the pontoon deck to the waterline.

3.8.4 Stability The floating dry dock and vessel to be drydocked shall be checked to ensure adequate system stability (vessel and dry dock). Longitudinal and transverse stability shall be considered, with the minimum allowable vessel KG controlling. At a minimum, the following five phases of the drydocking operation shall be checked (see Figure 3-1):

1. Phase 1: Floating dry dock at full submergence. The vessel is floating independently and is not yet in contact with the blocks.
2. Phase 2: Partial lift of the vessel. The vessel has contacted the blocks and has been lifted approximately half of its drydocking draft.
3. Phase 3: External waterline is just at top of keel blocks.
4. Phase 4: External waterline is just above the pontoon deck.
5. Phase 5: Floating dry dock is at normal operating draft. External waterline is below the pontoon deck.

The dry dock system (vessel and dry dock) shall maintain a minimum GM throughout the drydocking operation. The required minimum GM of the dry dock system (vessel and dry dock) varies with the certified rated capacity of the floating dock. The transverse GM of the dry dock system (vessel and dry dock), after all free surface corrections are made, shall not be less than 5.0 ft (1.525 m) for dry docks with a certified rated capacity of up to 10,000 long tons (10,200 metric tons). The GM may be reduced linearly for rated capacities over 10,000 long tons (10,200 metric tons) to a minimum of 3.28 ft (1.0 m) for a certified rated capacity of 50,000 long tons (51,000 metric tons) and above. Figure 3-2 shows the curve of the minimum required transverse GM as a function of the floating dry dock's lift capacity.

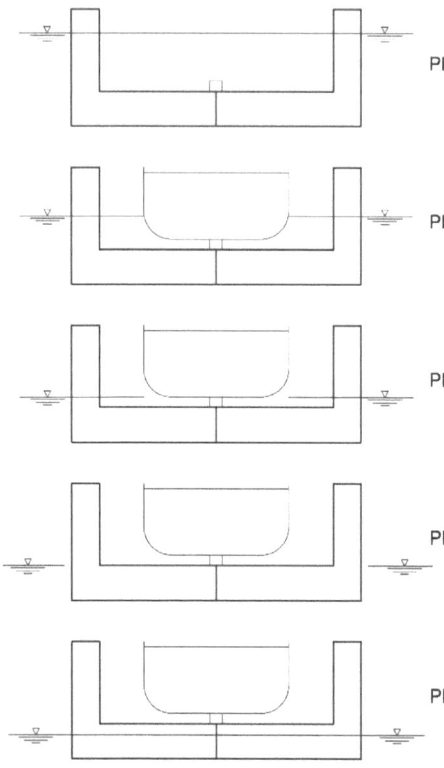

Figure 3-1. Phases of stability during drydocking operation.

3.8.5 Ballast Tank and Buoyancy Chamber Watertightness All ballast tanks and buoyancy chambers shall be watertight. Except for the ballast tank vents, no holes are permitted in ballast tanks. Utility penetrations must be sealed.

3.9 SPECIFIC REQUIREMENTS FOR GRAVING DOCKS

At this time, there are no specific requirements for graving docks.

3.10 SPECIFIC REQUIREMENTS FOR MARINE RAILWAYS

3.10.1 Line, Grade, and Gauge Survey The shipyard shall establish the allowable tolerances for the marine railway line, grade, and gauge.

Conduct a detailed line, grade, and gauge survey of the rails (using surveyor's instruments) at least once every 5 years or if unusual wear of rail plates (or rails) and/or rollers (or wheels) is observed.

The results of the survey shall be tabulated and documented graphically. Rails out of alignment shall be adjusted to bring them into alignment.

3.10.2 Chain Inspection Periodicity For marine railways that operate with a system of hauling chains and backing chains, inspect and document findings for the following:

1. If the hauling chain is over 5 years old and every 2 years thereafter, and/or if the chain shows visible signs of wear, gauge the hauling chain, measuring the barrel diameters of each side of the link, the double grip dimension, and the pitch distance over five links.
2. If the backing chain is over 5 years old and every 2 years thereafter, and/or if the chain shows visible signs of wear, gauge the backing chain, measuring the barrel diameters of each side of the link and the double grip dimension.

3.10.3 Wire Rope For marine railways that utilize a wire rope, the wire rope shall be replaced every 10 years.

3.10.4 Rollers The shipyard shall

1. Establish the ideal roller geometry and allowable fabrication tolerances for new rollers.
2. Establish allowable roller geometry for worn rollers; consideration shall be given for the possibility that a new roller or series of new rollers may be placed into service with in-service/worn rollers, without creating adverse consequences.
3. Establish an official shipyard roller drawing with revision control. The drawing should show the strict dimensional fabrication tolerances required for new rollers. The drawing should also show allowable roller geometry in the worn condition (i.e., acceptable wear). The drawing will serve three purposes
   (a) As the official roller fabrication drawing for new rollers,
   (b) As the basis for acceptance criteria prior to putting new rollers into service, and
   (c) As the basis for acceptable wear, which will be used for certification inspections by the certifying agency.
4. Establish acceptance criteria for new rollers.

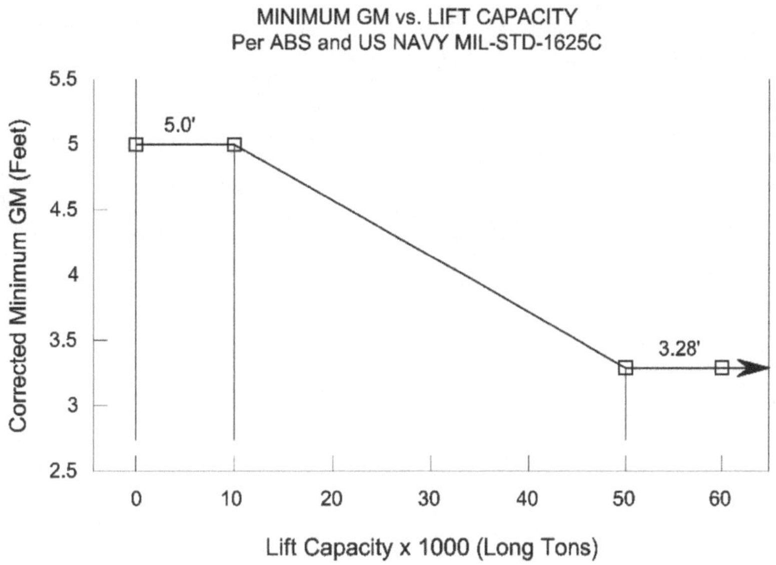

Figure 3-2. Minimum required GM versus the dry dock's certified rated capacity.

5. Check geometry and sign off prior to placing into service.
6. Establish a roller replacement implementation strategy. New rollers or roller trucks should not be placed immediately adjacent to worn rollers currently in service unless those rollers are found to be within allowable geometric tolerances.

3.11 SPECIFIC REQUIREMENTS FOR VERTICAL LIFTS

For vertical lifts that utilize a wire rope, the wire rope shall be replaced every 10 years.

CHAPTER 4
CONDITION ASSESSMENT REQUIREMENTS FOR FLOATING DRY DOCKS

4.1 DOCUMENT REVIEW

Documents governing the original design, construction, rehabilitation, alteration, and repair of the drydocking facility shall be reviewed. The design data required will vary, depending on the type of floating dock being evaluated. In general, the document review shall include

1. General arrangement of the floating dock;
2. Environmental data, including seismic, water, and wind data;
3. Geotechnical data (for moorings);
4. Design calculations;
5. Structural drawings and specifications;
6. Floating dry dock light weight survey;
7. Floating dry dock stability data;
8. Floating dry dock mooring design structural drawings and specifications;
9. Mechanical drawings and specifications;
10. Electrical drawings and specifications;
11. Instrumentation, control, and monitoring systems;
12. Computer-controlled ballast system (if applicable);
13. Deflection monitoring system (if applicable);
14. Floating dry dock translation system (if applicable);
15. Operating manual/procedures; and
16. Hydrographic data.

4.2 MATERIAL CONDITION SURVEY

A material condition survey is required to establish the existing condition of the drydocking facility. The condition of the structural steel, concrete, and/or timber shall be assessed in accordance with industry standards and documented in the certification inspection checklists. In general, the material condition survey shall include

1. Visual survey of internal ballast tanks and sea chests,
2. Visual survey of internal nonballasting compartments,
3. Visual survey of external portions of the floating dry dock
   (a) Pontoon deck,
   (b) Wingwalls, and
   (c) Miscellaneous external components.
4. Visual survey of the floating dry dock mooring system,
5. Visual survey of floating dry dock fenders and vessel moorings,
6. Floating dry dock translation system (if applicable),
7. Visual survey of vessel support blocks,
8. Underwater survey, and
9. Cathodic protection system (if applicable).

4.2.1 Internal Ballast Tanks and Sea Chests Inspect 100% of all ballast tanks and sea chests. Remove mud in accordance with the certifying agent's direction to accomplish a representative sampling.

Inspect all structural members for condition of protective coating, extent of corrosion, buckling, cracking, or other signs of distress. Inspect the bottom plate and plate stiffeners for signs that the floating dry dock has grounded (pushed-up plate, buckled stiffeners, and/or frames). Inspect each sea chest for condition of protective coating, extent of corrosion, buckling, cracking, or other signs of distress. Inspect the side shell plate and plate stiffeners for signs of external impact damage (pushed-in plate, buckled stiffeners, and/or frames). Inspect the ballast tank vents to ensure that there are no corrosion holes or other deficiencies.

4.2.2 Internal Nonballasting Compartments Inspect 100% of all nonballasting compartments, including void spaces, machinery spaces, and safety deck areas. Inspect all structural members for condition of protective coating, extent of corrosion, buckling, cracking, or other signs of distress. Inspect the side shell plate and plate stiffeners for signs of external impact damage (pushed-in plate, buckled stiffeners, and/or frames).

4.2.3 External Portions of the Floating Dry Dock Inspect all above-water external areas of the dry dock. These areas include the pontoon deck, wingwall weather deck, wingwall side shells, pontoon side shells, end bulkheads, aprons, fenders, and so forth.

4.2.3.1 Pontoon Deck Inspect the pontoon deck for severe "washboarding" of the deck (dishing of the deck plate between stiffeners). An example of washboarding is shown in Figure 4-1. The pontoon deck is usually one of the first areas to show deterioration. Heavy corrosion is not always evident, however, because rust scale is being continuously worn away by high traffic and heavy wear. This can significantly diminish the transverse bending strength of the pontoon.

4.2.3.2 Wingwalls Inspect the wingwall weather decks. The wingwall weather decks are the top structural flange of the floating dry dock which contributes to the dry dock's longitudinal strength. This is an extremely important structural member and shall not have any cracks, buckles, holes, or severe deterioration. Inspect the wingwall weather deck and the upper portions of the wingwalls for any large buckles that progress across the entire width of the deck and down the wingwall sides. Inspect for cracks in plates. Buckling or cracking in the upper portions of the wingwall is an indication of longitudinal overstress.

Inspect the wingwalls. The longitudinal strength of the dry dock is severely impaired by buckling or cracking in the upper regions of the wingwall and also in the lower region of the wingwall at the gaps if the pontoons are sectional. If buckling or cracking is found in these portions of the wingwall, the dry dock shall not be used until the situation can be investigated more thoroughly and repairs are made, if necessary.

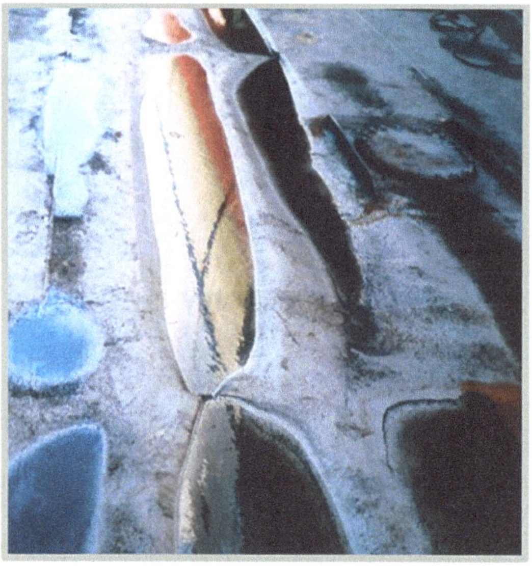

Figure 4-1. Washboarding of pontoon deck.
Source: Courtesy of Heger Dry Dock, Holliston, Massachusetts.

4.2.4 Floating Dry Dock Mooring System Inspect the floating dry dock's mooring system for signs of corrosion and wear.

4.2.5 Floating Dry Dock Fenders and Vessel Moorings Inspect and document the condition of foundation plating around all cleats, bollards, winches, capstans, and other components, especially where the component fastens to the deck, which is, in general, an area highly susceptible to corrosion. Inspect the condition of timber fenders and attachments to wingwalls, especially at the interface, where wood tends to rot and steel tends to corrode.

4.2.6 Floating Dry Dock Translation System (if applicable) Inspect the floating dry dock's translation system for signs of corrosion and wear.

4.2.7 Vessel Support Blocks Visually assess the condition of the vessel support blocking system. Inspect the timber for excessive crushing, warping, cracking, checking, rotting, or wear from fasteners. Inspect for loss of contact at edges. If the blocks have concrete bases, inspect the concrete for cracking, spalling, and exposed rebar. If the blocks have steel bases, inspect the steel for loss of metal thickness by corrosion, cracks, cracked welds, and other signs of defects. Inspect the condition of fastenings. All timber shall be adequately secured to prevent floating when the dock is submerged.

If the dry dock is equipped with hauling side blocks, inspect the hauling block hardware. The hauling block hardware typically includes chains, sheaves, cranks, slides, and pawls.

4.2.8 Cathodic Protection System Inspect the cathodic protection system (if applicable).

4.3 OPERATIONAL TESTS

In addition to the material condition survey, dry dock operational tests are required to establish the existing condition and function of the dry dock. In general, the operational tests shall include

1. Dry Dock Operation Test,
2. Submergence Test,
3. Light Weight Survey,
4. Translation System Test (if applicable), and
5. Verification of Instrumentation and Controls.

4.3.1 Dry Dock Operation Test The surveyor shall witness one complete cycling of the floating dry dock from operating freeboard to maximum submergence depth, and back to operating freeboard. The operational test shall include the drydocking or undocking of a vessel. The size and displacement weight of the vessel in relation to the operational limitations and capacity of the dry dock shall be at the discretion of the certifying authority. The dry dock's mechanical and electrical components and associated control systems shall be operated by the shipyard and witnessed by the surveyor. The operational test shall include the following:

1. Ballast pumps: Inspect for proper control from control panel and any abnormal vibration;
2. Ballast system valves: Inspect for proper control from control panel and that valve position indicator gauges or lights work properly;
3. Tank level indicator gauges;
4. Draft level indicator gauges and draft boards;
5. Primary power source;
6. Emergency backup power (if applicable): Demonstrate the maximum number of ballast pumps that are capable of being operated on the backup power system. The backup power system shall be capable of running the floating dry dock controls system, alarms, emergency lighting, and a sufficient number of ballast pumps to maintain safe operation (minimum of two pumps), albeit at a slower rate, simultaneously;
7. Capstans;
8. Winches;
9. Vessel centering system (if so equipped);
10. Longitudinal deflection monitoring system (requirement for docks 300 ft (91 m) and longer);
11. Alarms;
12. Lighting;
13. Emergency lighting; and
14. Fire protection system.

4.3.2 Submergence Test A submergence test shall be conducted by the shipyard and witnessed by the surveyor. Ballast the dry dock down to maximum submergence and record the floating dry dock drafts. In dry docks designed with a safety deck, the submergence shall cease at the maximum design draft with the flood valves open. If the dry dock cannot achieve its design draft, the reason shall be noted (e.g., insufficient water depth at site, incorrect vent tube lengths.).

Inspect the safety deck and the cross-over tunnel (if applicable) when the dock is fully submerged. At this time, the air in the top of the ballast tanks is pressurized. Any leaks in the safety deck plate, non-ballasting compartments, or packing glands will be evident by hissing air or leaking water.

Close the flood valves and raise the dry dock 5 ft (1.52 m). Record the floating dry dock drafts. After a minimum of 30 min, record the floating dry dock drafts again, just prior to deballasting the floating dry dock. Any change in drafts, indicating leakage, shall be documented by the surveyor. If leaks are detected, the source of the leak shall be identified. To accomplish this, further testing may be required to isolate the source of the leaks.

Deballast the dock using all the main ballast pumps. Record the time to ballast (submerge) and deballast (raise) the floating dry dock, with starting and stopping dock drafts noted.

4.3.3 Lightweight Survey With no vessel on the dry dock, dewater all ballast tanks to their minimum ballast water level using only the ballast pumps. Record the depth of the remaining residual water. The use of portable pumps to remove the residual water is not permitted.

With the ballast tanks deballasted, measure the pontoon freeboard at the lowest point on the pontoon deck (usually at the inside corner of the wingwall) to the waterline. The buoyancy available to lift the vessel is represented by the volume of the floating dry dock's pontoon above the waterline to the pontoon's operating freeboard.

4.3.4 Translation System Test (if applicable) The surveyor shall witness one complete cycling of the floating dry dock's translation system. Inspect for signs of corrosion and wear.

4.3.5 Verification of Instrumentation and Controls Verify that all instrumentation and controls are in good working order. If, during the operational test, the water level indicator gauges, ammeter gauges, valve position indicators, or other instrumentation do not reflect the existing condition, the errant gauges shall be recalibrated and checked by comparing the physical measurement to the indication on the gauge.

The surveyor shall verify the accuracy of the deflection monitoring system (as applicable) by determining the dry dock's actual deflection with an independent measuring system and comparing the results with the dry dock's system.

4.4 DRY DOCK CALCULATIONS

Establish, verify, or adjust the capacity, stability, and operational limitations of the floating dock in its present condition. The dry dock evaluation shall include the following:

1. Light dock weight determination,
2. Buoyant capacity,
3. Stability,
4. Transverse bending,
5. Longitudinal bending and deflection limits,
6. Local strength analysis,
7. Block loading analysis,
8. Hydrostatic properties of the hull,
9. Hydrostatic head pressures on internal bulkheads,
10. List and list trims,
11. Mooring analysis,
12. Corrosion criteria, and.
13. Calculation of maximum sill load (for vessel transfer systems).

4.4.1 Light Weight Determination If the original light weight survey, floating dry dock weight, and/or KG of the floating dry dock are unavailable, or if the floating dry dock's weight and/or KG have been modified since the original calculations were performed, conduct a lightweight survey to determine the current (existing) floating dry dock weight and KG.

4.4.2 Buoyant Capacity The buoyant capacity of the dry dock is calculated by subtracting the floating dry dock's displacement at minimum draft from the floating dry dock's displacement at the pontoon operating freeboard, as can be seen in Figure 4-2.

The operating draft is calculated by subtracting the operating pontoon freeboard from the pontoon depth at the location of the freeboard measurement (Figure 4-3).

4.4.3 Stability Calculate the minimum stability phase and develop a vessel KG versus vessel displacement weight curve for the floating dry dock for that minimum stability phase. The vessel KG versus vessel displacement weight curve is a plot of the maximum allowable vessel KG (the distance the vessel's center of gravity is above the vessel's keel, adjusted for free liquids onboard the vessel) versus the vessel's displacement weight at time of drydocking, which results in the minimum allowable GM for the floating dry dock and vessel combination. Figure 4-4 shows a sample vessel KG versus vessel displacement weight curve.

This curve is used by the Dockmaster and/or Dry Dock Engineer to verify adequate stability of the floating dry dock and vessel combination when drydocking any vessel. If the plot of the vessel's drydocking displacement weight and its KG intersect on or below the line, the floating dry dock has sufficient stability to lift the vessel and maintain the minimum GM for the floating dry dock and vessel system. If the plot of the vessel's drydocking displacement weight and its KG intersect above the line, the floating dry dock does not have sufficient stability to lift the vessel and the GM for the floating dry dock and vessel system will be below the minimum required indicating an unstable condition.

4.4.4 Transverse Bending Calculate at least four separate loading conditions when analyzing transverse strength

1. Maximum vessel load at the dry dock centerline and external water at top of keel blocks (Phase 3),
2. Maximum vessel load at the dry dock centerline and at operating draft (Phase 5),

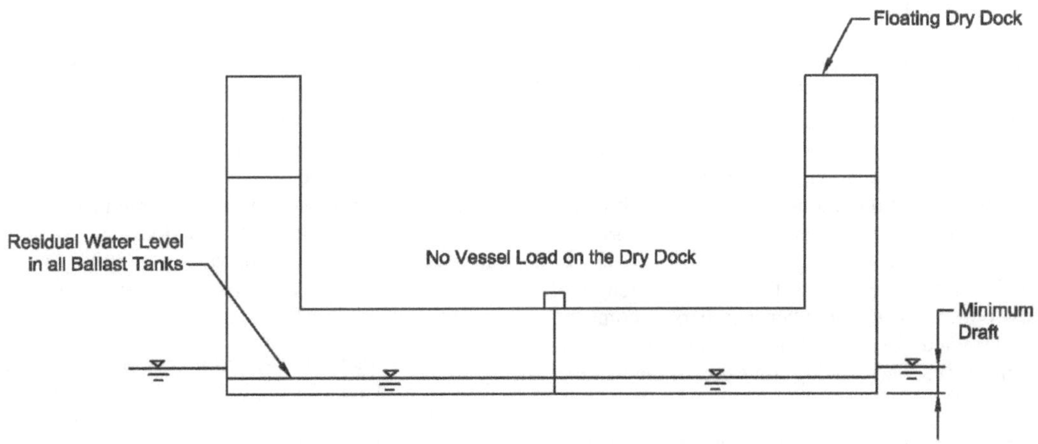

Figure 4-2. Minimum dry dock draft.

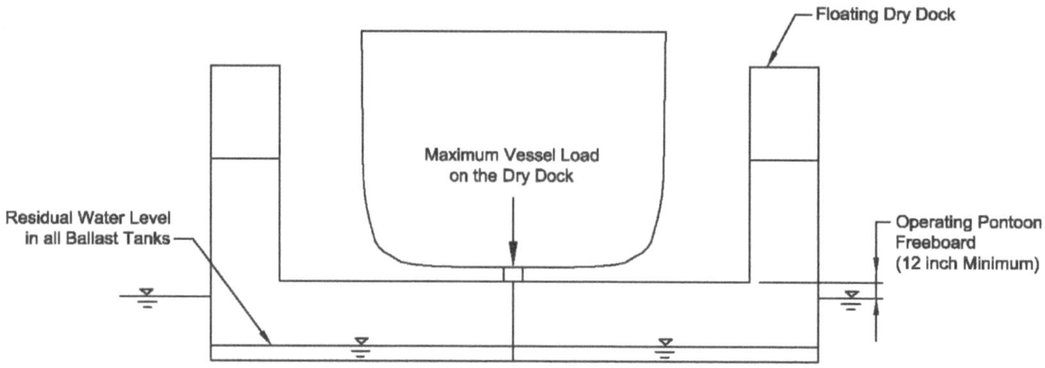

Figure 4-3. Dry dock operating draft.

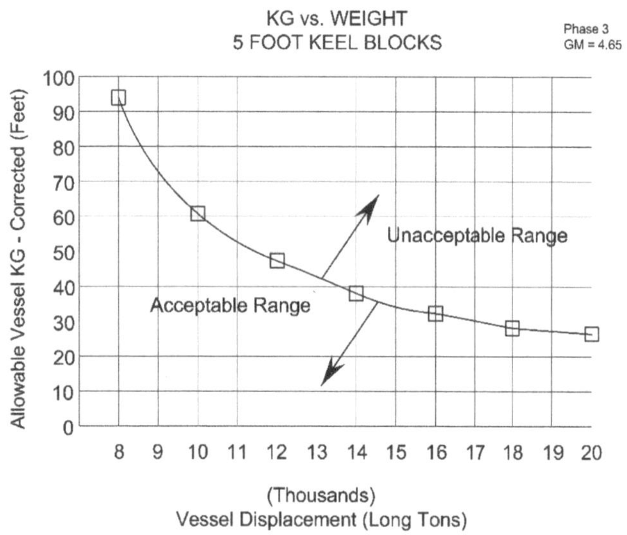

Figure 4-4. Sample vessel KG versus vessel displacement weight curve.

3. Partial vessel load at the dry dock centerline and internal water at bottom of wingwalls causing maximum head condition, and
4. Reverse bending owing to the dewatering of an unloaded length of the dock. This will occur when only a portion of the length of a ballast tank has vessel load on it.

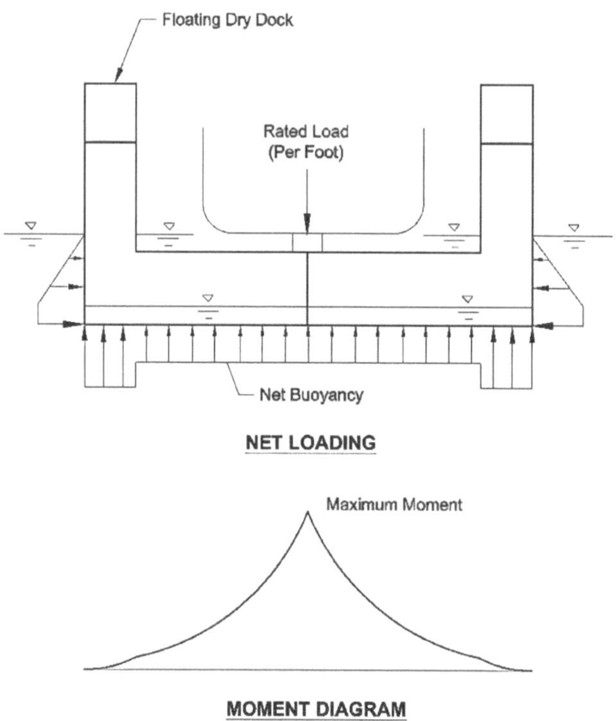

Figure 4-5. Maximum transverse bending condition.

The transverse strength of the dry dock is provided by the transverse bulkheads (watertight and nonwatertight) and/or transverse trusses in the pontoon. The pontoon structure shall distribute the concentrated load of the ship along the floating dry dock's centerline to the buoyant support of the water over its entire width by its transverse strength. Combine the transverse bending stresses with stresses caused by local conditions such as hydrostatic pressure on the hull, where applicable. If the section modulus of the transverse frame varies across the width of the floating dry dock, the maximum transverse bending stress shall be calculated for each different cross section.

When determining allowable stresses, the buckling strength of the pontoon deck, bottom plate, and other members in compression shall be determined for each different panel size, plate thickness, or member size. The existing condition of the member shall be used.

4.4.4.1 Transverse Bending with Water at Top of Keel Blocks Calculate the bending moment at the point when the exterior water is at the top of the keel blocks as shown in Figure 4-5. During the drydocking evolution, the dry dock/ship system passes through a condition in which the water level outside the dry dock is at the top of the keel blocks. This condition produces the maximum transverse bending in the floating dry dock. At this time, 100% of the vessel weight is on the floating dry dock while the pontoon and the submerged section of the wingwall provide buoyant lift.

The submerged section of the wing provides additional buoyancy farther away from the dry dock centerline, which increases the bending moment. For this case, 100% of the rated load per foot is assumed to act on the keel blocks at the transverse centerline.

4.4.4.2 Transverse Bending at Operating Draft Calculate transverse bending stresses when the dry dock is at its operating draft. In cases in which the rated load per foot exceeds the buoyant capacity of the dry dock at the operating draft, the additional uplift is supported through shear in the wingwalls and longitudinal

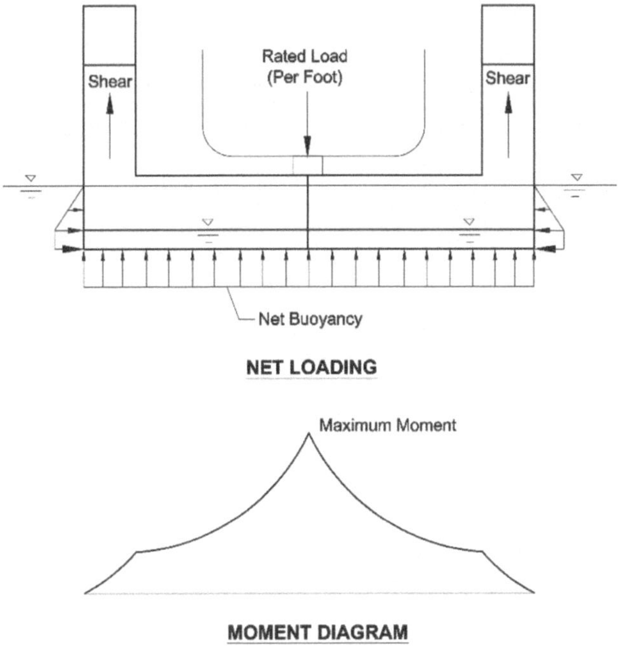

Figure 4-6. Transverse bending at operating draft.

bulkheads. This additional uplift increases the transverse bending moment (Figure 4-6).

4.4.4.3 Partial Load, Maximum Head Condition Analyze transverse bending for partial load, maximum head condition. The partial load, maximum head condition is the point at which the maximum hydrostatic head occurs on the shell of the floating dry dock. Although the load at the centerline is only a portion of the total vessel load and the bending moment is not as great as in the first case, the bending stresses, when combined with the local stresses caused by hydrostatic head pressure, may control. The partial load, maximum head condition occurs when the internal water level has reached the base of the wings (Figure 4-7).

The magnitude of the hydrostatic head, which develops at this point, is a function of the vessel's draft and beam. The deeper the draft and wider the beam of the vessel being docked, the higher the maximum hydrostatic head.

The volume of water that is pumped out of the wingwalls (V1) after the vessel's keel contacts the keel blocks is equal to the volume at which the vessel and dry dock wings rise out of the water during that pumping (V2). With this volume known, along with the length and beam of the ship and dock wingwalls, the rise of the ship and dock can be calculated (Figure 4-8).

The resulting depth of water over the pontoon deck is the maximum head on the dry dock for that particular vessel. The displaced volume of the portion of ship that has been lifted out of the water (V2 Ship) is the weight of the vessel on the keel blocks for this condition.

4.4.4.4 Reverse Bending Analyze the reverse transverse bending condition. This condition occurs wherever there is a dewatered ballast tank or nonballasting tank providing uplift in an area where there is no vessel weight pushing down, such as an unloaded portion of a partially loaded compartment (Figure 4-9).

The pontoon is held down at the wingwalls and buoyancy tends to bow it up, the reverse of a typical loading condition. In this case, the pontoon bottom plate is in compression and shall be analyzed for plate buckling.

4.4.4.5 Transverse Bending Limits Determine the dry dock's rated maximum allowable load per foot along the keel. With the dry dock supporting the maximum load per foot, transverse bending stresses for each of the aforementioned bending cases shall not exceed the established acceptable limits for each member.

4.4.5 Longitudinal Bending Calculations and Deflection Limits Calculate the longitudinal bending stresses in the dry dock for the shortest capacity vessel that will not exceed the dry dock's load per foot limitations. Analyze the condition in which a vessel much shorter than the floating dry dock with a weight per foot at the certified rated capacity produces an extreme longitudinal bending condition. This longitudinal bending condition results from the buoyant uplift force on the dry dock forward and aft of the vessel combined with the deadweight force of the vessel (Figure 4-10).

For dry docks longer than 300 ft (91 m), calculate the longitudinal deflection limits of the dry dock for various loading conditions such that the minimum dry dock deflection that causes the floating dry dock to reach working stress limits shall be established. Allowable buckling stresses in the wingwall weather deck, upper wingwall side shell, and pontoon bottom, if lower than normal bending stresses, shall control.

These calculations shall be used to set the operational longitudinal deflection limits.

4.4.6 Local Strength Analysis Calculations shall be performed to determine

1. Allowable hydrostatic pressure on hull,
2. Allowable hydrostatic pressure on internal bulkheads and safety deck,
3. Allowable vehicle loading on pontoon deck, aprons, access bridge, and so forth,
4. Mooring loads, and
5. Foundation loads.

4.4.7 Block Loading Analysis Perform calculations to determine that the vessel support blocks are stable and structurally adequate to support the floating dry dock's certified rated load per block. Calculations shall include

1. Allowable bearing pressure on timber times bearing area.
2. Structural strength of block base unit,
3. Local structural strength of floating dock structure directly supporting blocks,
4. Rated load per keel block. The rated load per keel block shall be the lesser of
   (a) Allowable bearing pressure on timber times bearing area, or
   (b) Structural strength of block base unit, or
   (c) Local structural strength of floating dry dock structure directly supporting blocks.

4.4.8 Hydrostatic Properties of the Hull Calculate and document the following hydrostatic properties of the dry dock hull:

1. Displacement of dock versus draft,
2. KB of floating dry dock versus draft,
3. KM of floating dry dock versus draft,
4. Transverse inertia of floating dry dock versus draft, and
5. Ballast tank tables or lift curves.

State the allowable hydrostatic head pressures on the floating dry dock's external shell (Figure 4-11).

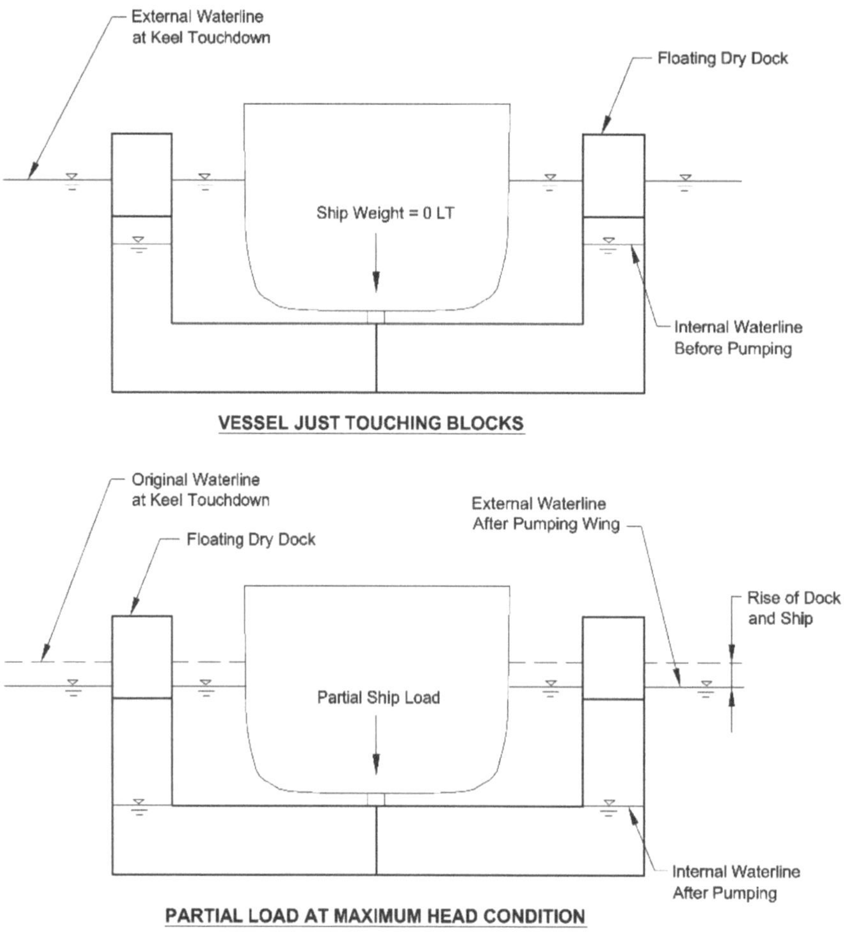

Figure 4-7. Partial load at maximum head condition.

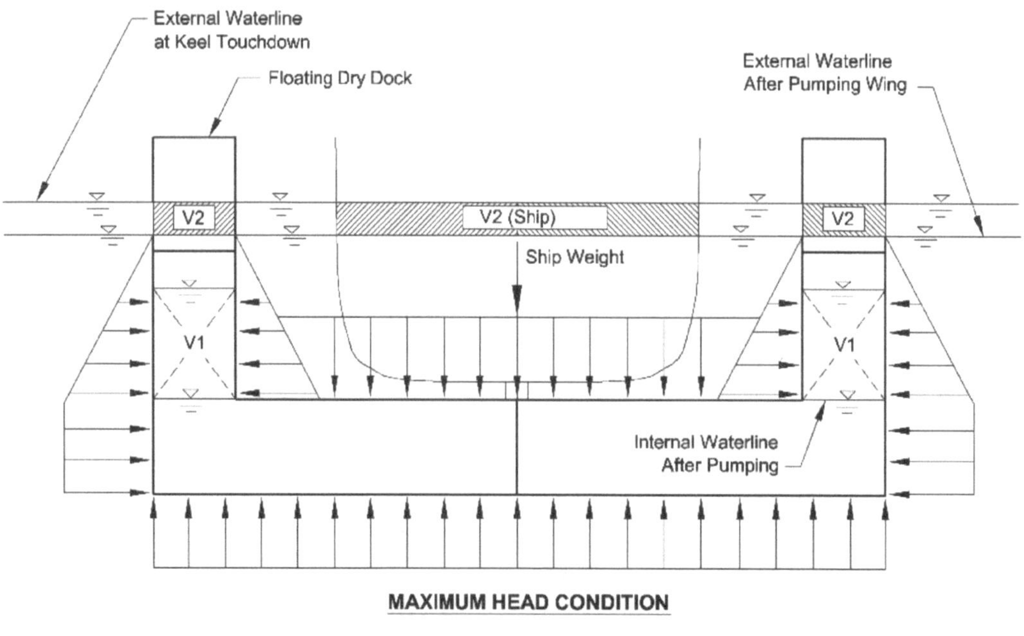

Figure 4-8. Maximum head condition.

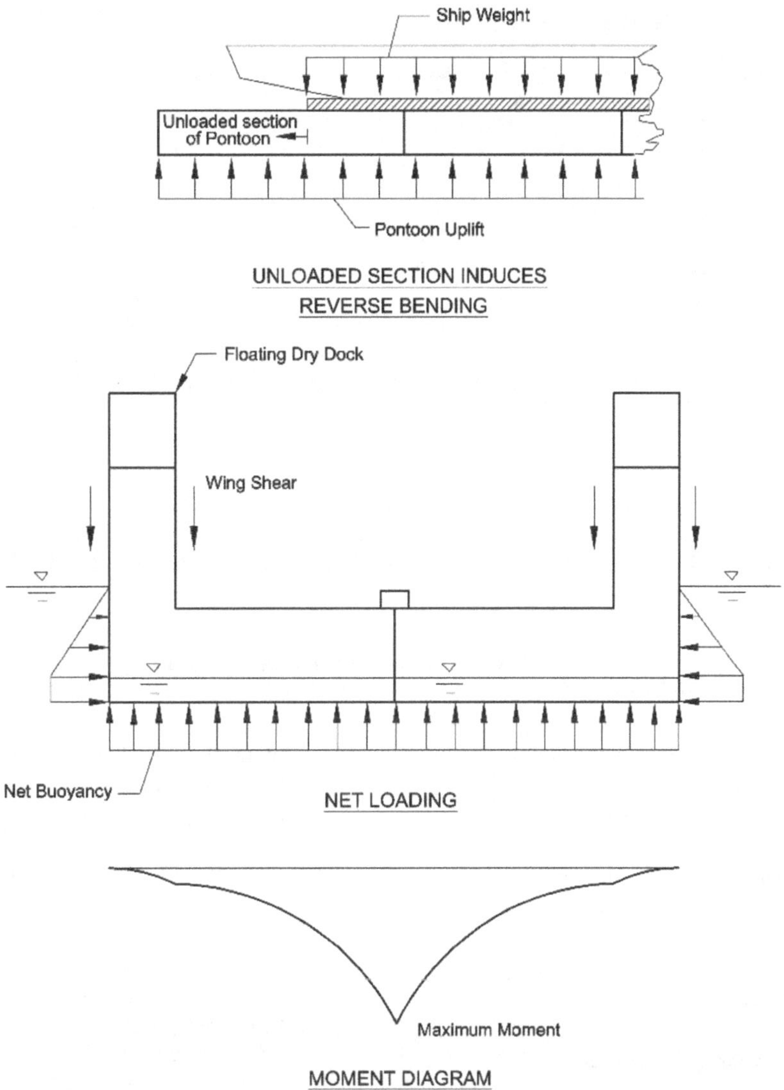

Figure 4-9. Reverse bending at unloaded section.

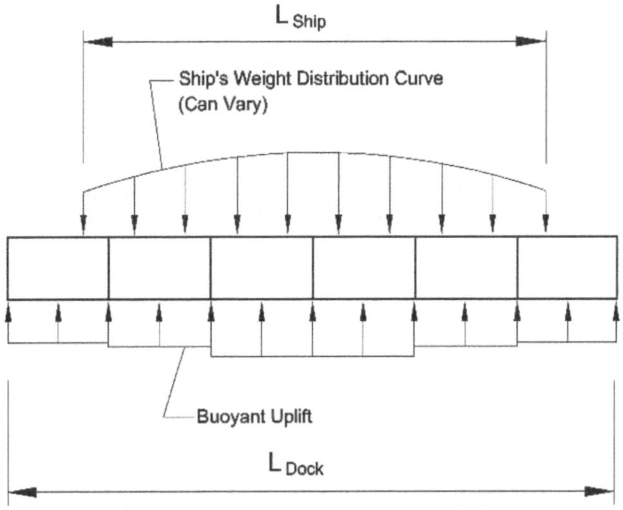

Figure 4-10. Longitudinal bending condition.

4.4.9 Hydrostatic Head Pressures on Internal Bulkheads Calculate the allowable hydrostatic head pressures of the floating dry dock's internal bulkheads.

4.4.10 List and Trim Limits Analyze and document the maximum list and trim as it relates to possible interference with the mooring system and other adjacent structures, or jeopardizes the stability of installed cranes, access bridges, and other equipment.

State the maximum list and trim angles for the dock such that the mooring system does not bind; the wingwall cranes or other structures are not in danger of falling or rolling; the dry dock will not contact bulkheads, piles, piers, and other structures; and access bridges are not endangered.

4.4.11 Mooring Analyses Analyze the dock's mooring system. The analysis shall establish

1. Maximum wind speed in combination with maximum current speed that the mooring system is capable of resisting.

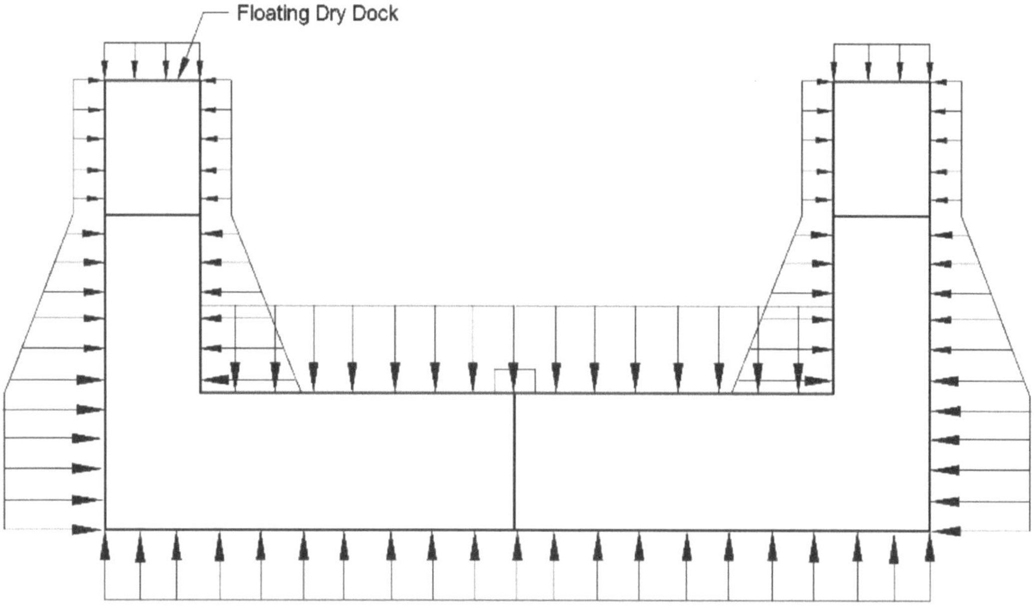

Figure 4-11. External head pressure diagram.

2. Maximum trim and list that the mooring system is capable of achieving without binding.
3. Maximum and minimum tide elevations.

4.4.11.1 Corrosion Criteria Provide a description of the corrosion criteria used in analyzing the floating dry dock's structure.

4.4.11.2 Calculation of Maximum Sill Load (if applicable) If applicable, determine the maximum sill load.

4.5 CONDITION ASSESSMENT REPORT

In addition to the requirements of Chapter 3 of this standard, include in the written Condition Assessment Report the items in Section 4.5.1.

4.5.1 Written Descriptions Provide a description of the following:

1. Floating dry dock ballast system. Include original design pump and flood times if known.
2. Floating dry dock ballast system controls.
3. Floating dry dock translation system (if applicable).
4. Floating dry dock blocking system(s).
5. The method of monitoring longitudinal deflection.

4.5.2 Calculations Include and provide a written summary of the calculations required by Chapter 4 of this standard.

4.5.3 Floating Dry Dock Capacity

4.5.3.1 Overall Lift Capacity State the overall lift capacity of the floating dry dock. The overall lift capacity shall be the lesser of

1. Buoyant capacity of the floating dry dock at the minimum pontoon deck freeboard, or
2. Rated load per foot capacity times the available keel blocking length, or
3. Vessel KG owing to stability requirements.

4.5.3.2 Rated Load per Foot Capacity State the rated load per foot capacity of the dock. The rated load per foot capacity shall be controlled by the lesser of

1. Transverse bending strength of the dock, or
2. Longitudinal bending strength of the dock, or
3. Keel block capacity, or
4. Local strength of the block support structure in pontoon.

4.5.3.3 Allowable Pontoon Deck and Apron Loading State the maximum allowable pontoon deck and apron loading.

CHAPTER 5
CONDITION ASSESSMENT REQUIREMENTS FOR GRAVING DOCKS

5.1 DOCUMENT REVIEW

Documents governing the original design, capacity, construction, rehabilitation, alteration, and repair of the drydocking facility shall be reviewed. The design data required will vary depending on the type of graving dock being evaluated. In general, the document review shall include

1. General arrangement of the graving dock,
2. Environmental data, including seismic, water, and wind data,
3. Geotechnical data,
4. Design calculations,
5. Stability data of closure gate (if applicable),
6. Structural drawings and specifications,
7. Mechanical drawings and specifications,
8. Electrical drawings and specifications,
9. Instrumentation, control, and monitoring systems,
10. Operating manual/procedures,
11. Hydrographic data, and
12. Concrete material test data, including but not limited to concrete cylinder breaks (compressive strength) and flexural strength tests (modulus of rupture).

5.2 MATERIAL CONDITION SURVEY

A material condition survey is required to establish the existing condition of the drydocking facility. The condition of the structural steel, concrete, and/or timber shall be assessed in accordance with industry standards and documented in the certification inspection checklists. In general, the material condition survey shall include

1. Visual survey of graving dock floor,
2. Grade survey of graving dock floor (if necessary),
3. Visual survey of graving dock walls,
4. Displacement survey of graving dock walls (if necessary),
5. Visual survey of graving dock sill, closure gate abutments, and entrance structures,
6. Visual survey of the pumphouse/pumpwell and other subsurface structures,
7. Visual survey of flooding/dewatering tunnels,
8. Visual survey of area surrounding graving dock,
9. Visual survey of the closure and intermediate gates,
10. Visual survey of graving dock fenders and vessel moorings,
11. Visual survey of vessel support blocks,
12. Underwater survey, and
13. Cathodic protection system (if applicable).

Depending on the results of the visual survey and the availability of design data, the survey may also include

1. Coring into the existing floor slab to obtain concrete cylinders for strength testing,
2. Coring into the existing floor slab to investigate the presence of voids,
3. Additional soil sampling and testing,
4. Visual inspection of the underdrain system via remote camera, and
5. Mooring hardware load test/anchor bolt pull test.

5.2.1 Graving Dock Floor Inspect the graving dock floor for depressions, cracks, spalling, leakage, or other types of degradation. The size and location of major cracks and leaks shall be mapped and recorded for future reference. Evidence of upward displacement or settlement of the floor, or evidence of voids under the floor, shall be documented and investigated further to determine cause. For graving docks designed with a system to relieve hydrostatic pressure under the floor slab, verify the system is working as designed.

5.2.2 Grade Survey of Graving Dock Floor Measure the elevation of the floor slab and record findings for future reference. This information will be used in future surveys to determine whether the slab is rising or settling.

5.2.3 Graving Dock Walls Inspect the graving dock walls. Dry docks with concrete walls shall be inspected for cracks, spalling, leakage, movement, or other types of degradation. Map and record the size and location of major cracks and leaks for future reference. Dry docks with steel sheet-pile walls shall be inspected for corrosion, loss of metal thickness, split seams, bulging, buckling, or other types of degradation or failure. For graving docks designed with a system to relieve hydrostatic pressure behind the walls, verify the system is working as designed.

If there are visual indications that the basin sidewalls may have moved from their original location, survey the actual position of the walls and record findings for future reference. This information will be used in future surveys to determine whether the walls are continuing to move.

5.2.4 Graving Dock Sill, Closure Gate Abutments, and Entrance Structures Inspect the graving dock sill, closure gate abutments, entrance structures, and other structural elements for concrete cracks and spalls, leakage, movement, or other types of deterioration and/or signs of distress. Any steel sheet pile walls or steel structural members shall be inspected for corrosion, loss of metal thickness, split seams, bulging, buckling, or other types of degradation or failure.

5.2.5 Pumphouse/Pumpwell and other Subsurface Structures Inspect pumphouse/pumpwell and other subsurface structures, such as sluice gate operator pits and capstan pits. Inspect for concrete cracks and spalls, leakage, movement, or other types of degradation and/or distress. Steel structural members shall be inspected for cracks, buckling, severe corrosion, and other signs of deterioration/distress. Map and record the extent of the deficiencies for future reference. Inspect housed mechanical and electrical components, such as dewatering pumps, valves, sluice gate operators, and document any significant deterioration or signs of distress that may affect capacity and/or reliability.

5.2.6 Flooding/Dewatering Tunnels Inspect the flooding/dewatering tunnels for cracks, spalling, leakage, or other types of degradation. Map and record the size and location of major cracks and leaks for future reference. Inspect any mechanical components, such as sluice gates and stop logs, and document any significant deterioration or signs of distress that may affect capacity and/or reliability.

5.2.7 Areas Surrounding Graving Dock Inspect the area surrounding the graving dock for signs of settlement indicating loss of material.

5.2.8 Closure and Intermediate Gates For closure gates that have ballast tanks, inspect 100% of all of the ballast tanks and machinery spaces. The material condition of structural elements can vary greatly from ballast tank to ballast tank or even within a single ballast tank, owing to varying environmental conditions or past renovation projects that repaired some areas but not others. Therefore, it is required to visually inspect all members in every ballast tank. For closure gates without ballast tanks, inspect both sides of the closure gate.

Inspect all structural members for condition of protective coating, buckling, cracking, extent of corrosion, condition of fasteners, or other signs of distress. Remove silt or sediment from the ballast tanks prior to the inspection to provide access to 100% of the structure.

For closure gates outfitted with half-round pipe fenders welded to the external hull plating, inspect the fenders and hull plating for trapped water. Nonwatertight half-round fenders are prone to accelerated corrosion of the fender and hull plating behind the fenders. Inspect the side shell plate and plate stiffeners for signs of external impact damage (pushed-in plate, buckled stiffeners, and/or frames).

Assess the effectiveness of the closure gate seals and general fit of the gate. Inspect for marine borer damage. Inspect the condition of the seal connections. Document the size and location of all leaks (external, between tanks, through valves, packing glands, etc.)

5.2.9 Graving Dock Fenders and Vessel Moorings Inspect the condition of foundation plating for dock fenders and mooring hardware, such as cleats, bollards, winches, and capstans, especially where the component fastens to the deck or other structural surface, which is, in general, an area highly susceptible to corrosion.

5.2.10 Vessel Support Blocks Visually assess the condition of the vessel support blocking system. Inspect the timber for excessive crushing, warping, cracking, checking, rotting, or wear from fasteners. Check for loss of contact at edges. If the blocks have concrete bases, inspect the concrete for cracking, spalling, and exposed rebar. If the blocks have steel bases, inspect the steel for loss of metal thickness by corrosion, cracks, cracked welds, and other defects. Inspect the condition of fastenings. All timber shall be adequately secured to prevent floating when the dock is flooded.

If the dock is equipped with hauling side blocks, inspect the hauling block hardware. The hauling block hardware typically includes chains, sheaves, cranks, slides, and pawls.

5.2.11 Cathodic Protection System Inspect the cathodic protection system (if applicable).

5.3 OPERATIONAL TESTS

In addition to the material condition survey, dry dock operational tests are required to establish the existing condition and function of the dry dock. In general, the operational tests shall include (1) dry dock operation test and (2) verification of instrumentation and controls.

5.3.1 Dry Dock Operation Test The surveyor shall witness one complete cycling of the graving dry dock and closure gate. The operational test shall include the drydocking or undocking of a vessel. The size and displacement weight of the vessel in relation to the operational limitations and capacity of the dry dock shall be at the discretion of the certifying authority. The dry dock's mechanical and electrical components and associated control systems shall be operated by the shipyard and witnessed by the surveyor. The operational test shall include the following:

1. Dry dock dewatering, stripping, and superflooding (if applicable) pumps: check for proper control from control panel, abnormal vibration;
2. For closure gates designed to be ballasted and deballasted, inspect the ballast system of valves and pumps and the ballast control equipment;
3. For gates designed to swing in and out of position or lowered on hinges, inspect the operating mechanisms;
4. Dry dock flooding and dewatering system valves, stripping system valves, and sluice gates: check for proper control from control panel, valve position indicator gauges, or lights working properly;
5. Water level indicator gauges;
6. Capstans;
7. Winches;
8. Vessel centering system (if so equipped);
9. Control systems;
10. Water level sensors/high water alarms;
11. Lighting;
12. Emergency lighting;
13. Fire protection system; and
14. Cathodic protection.

5.3.2 Verification of Calibration of Instruments If, during the operational test, the water level indicator gauges, ammeter gauges, valve position indicators, or other instrumentation do not reflect the actual condition, the errant gauges shall be recalibrated and checked by comparing the physical measurement to the indication on the gauges.

5.4 DRY DOCK CALCULATIONS

Establish, verify, or adjust the capacity and the operational limitations of the graving dock in its present condition. The dry dock evaluation shall include the following:

1. Geotechnical calculations,
2. Loading on floor,
3. Loading on walls,
4. Loading on dry dock sill,
5. Uplift stability,
6. Structural strength of the closure gate,
7. Stability and hydrostatic properties of closure gate,
8. Light weight of closure gate determination,

9. Ballast required to seat closure gate,
10. Hydrostatic head pressure on closure gate,
11. Hydrostatic head pressure on internal bulkheads of the closure gate,
12. Vulnerabilities assessment with the closure gate removed,
13. Block loading calculations, and
14. Corrosion criteria.

5.4.1 Geotechnical Calculations Establish the geotechnical properties of the soils below the floor slab and behind the walls. These geotechnical properties shall be used in the structural capacity and stability calculations.

5.4.2 Loading on Floor Analyze critical loading combinations and determine the resulting stresses and factors of safety. If allowable block loading varies with location or blocking is limited to certain reinforced areas, these locations and load limits shall be documented.

5.4.3 Loading on Walls Analyze critical loading combinations and determine the resulting stresses and factors of safety.

5.4.4 Loading on Dry Dock Sill Analyze the critical loading combinations for the bearing of the closure gate on the dry dock sill and determine the resulting stresses, including gate seat shear stresses and factors of safety. If the dry dock has superflood capability, the analysis shall address this load scenario and determine the resulting stresses on the reverse gate seat.

5.4.5 Uplift Stability Calculations shall be provided that demonstrate that the uplift resistance furnished by the graving dock is greater than the uplift force caused by hydrostatic pressures (Figure 5-1).

5.4.6 Structural Strength of Closure Gate Calculate the structural strength of the closure gate.

5.4.7 Stability of the Closure Gate For closure gates that are designed to be floated in and out of position, calculate the stability of the gate when afloat. Calculate the minimum water levels required in the gate's ballast tanks to maintain adequate stability.

5.4.8 Light Weight Determination of Closure Gate For dry docks designed with a floating caisson gate, document the weight and center of gravity of the gate. If the original light weight survey, caisson gate weight, and/or vertical center of gravity (KG) of the caisson gate are unavailable, or if the caisson gate's weight and/or KG have been modified since the original calculations were performed, conduct a light weight survey to determine the current (existing) caisson gate weight and KG.

5.4.9 Ballast Required to Seat Closure Gate For gates that are designed to be floated in and out of position, determine the minimum amount of ballast water required to keep the gate seated. The light weight of the gate combined with the weight of the ballast water shall be greater than the buoyant force on the gate during the highest anticipated tide (e.g., storm condition) and factoring in design waves.

5.4.10 Hydrostatic Head Pressures on the Closure Gate Calculate the maximum tide level that the gate can resist with an empty dry dock basin.

5.4.11 Hydrostatic Head Pressures on Internal Bulkheads of the Closure Gate For closure gates that have internal ballast tanks, calculate the allowable hydrostatic head pressures of the closure gate's internal bulkheads.

5.4.12 Vulnerabilities Assessment with the Closure Gate Removed Conduct an assessment of vulnerabilities to the dry dock when the gate is removed during drydocking evolutions or when the closure gate is undergoing off-site maintenance. Determine the maximum allowable tide (water) level factoring in high tides, storm surge, and design waves; assess whether utility service galleries, subsurface capstan, and sluice gate pits, pumpwells, electrical and mechanical service vaults, and so forth, are vulnerable to flooding. For partially relieved dry docks in which the pressure-relief/drainage pump system is shut down for an extended period of time, factor in high groundwater effects. Document findings and any recommendations for facility improvements and/or preventive measures to minimize vulnerabilities when the closure gate is removed.

5.4.13 Block Loading Analysis Perform calculations to determine that the vessel support blocks are stable and structurally adequate to support the vessel. Calculations shall include

1. Allowable bearing pressure on timber times bearing area,
2. Structural strength of block base unit,

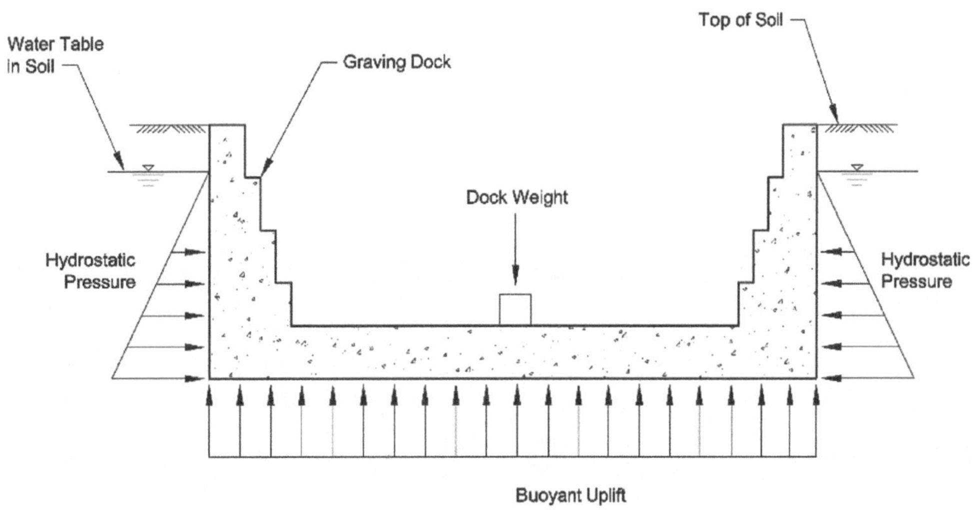

Figure 5-1. Hydrostatic forces on a graving dock.

3. Local structural strength of the dry dock structure directly supporting blocks,
4. State the rated load per keel block. The rated load per keel block shall be the lesser of
   (a) Allowable bearing pressure on timber times bearing area, or
   (b) Structural strength of block base unit, or
   (c) Local structural strength of the dry dock structure directly supporting blocks.
5. State the rated load per side block. The rated load per side block shall be the lesser of
   (a) Allowable bearing pressure on timber times bearing area, or
   (b) Structural strength of block base unit, or
   (c) Local structural strength of dry dock structure directly supporting blocks.

5.4.14 Corrosion Criteria Provide a description of the corrosion criteria used in analyzing the graving dock's structure.

5.5 CONDITION ASSESSMENT REPORT

In addition to the requirements of Chapter 3 of this standard, include in the written Condition Assessment Report the following items.

5.5.1 Written Descriptions Provide a description of the following:

1. Graving dock's flooding system,
2. Graving dock's dewatering system,
3. Graving dock's seepage water removal system,
4. Graving dock's underdrain system (if applicable),
5. Closure gate ballast system and operation (if applicable), and
6. Superflooding system (if applicable).

5.5.2 Calculations Include and provide a written summary of the calculations required by Chapter 5 of this standard.

5.5.3 Overall Capacity State the overall capacity of the graving dock. The overall capacity shall be the rated load per foot capacity multiplied by the available keel blocking length.

5.5.4 Rated Load per Foot Capacity State the rated load per foot capacity of the graving dock. The rated load per foot capacity shall be controlled by the lesser of

1. Strength of the dock floor slab, or
2. Strength of the floor slab supporting piles (if applicable), or
3. Strength of the dock floor subgrade soil, or
4. Keel block capacity.

CHAPTER 6
CONDITION ASSESSMENT REQUIREMENTS FOR MARINE RAILWAYS

6.1 DOCUMENT REVIEW

Documents governing the original design, capacity, construction, rehabilitation, alteration, and repair of the drydocking facility shall be reviewed. The design data required will vary depending on the type of marine railway being evaluated. In general, the document review shall include

1. General arrangement of the marine railway,
2. Environmental data, including seismic, water, and wind data,
3. Geotechnical data,
4. Design calculations,
5. Structural drawings and specifications,
6. Line, grade, and gauge survey,
7. Roller drawings and specifications,
8. Hauling and backing chain (or wire rope) drawings and specifications,
9. Mechanical drawings and specifications,
10. Electrical drawings and specifications,
11. Instrumentation, control, and monitoring systems,
12. Operating manual/procedures, and
13. Hydrographic data.

6.2 MATERIAL CONDITION SURVEY

A material condition survey is required to establish the existing condition of the drydocking facility. The condition of the structural steel, concrete, and/or timber shall be assessed in accordance with industry standards and documented in the certification inspection checklists. In general, the material condition survey shall include

1. Visual survey of track foundation,
2. Visual survey of tracks and rails
   (a) Concrete track,
   (b) Steel track, and
   (c) Timber track,
3. Visual survey of drydocking cradle,
4. Inspection of rollers or wheels,
5. Inspection of chains or wire rope,
   Hauling and backing chain,
   Wire rope,
6. Inspection of hauling machinery,
7. Visual survey of drydocking cradle fenders and vessel moorings,
8. Visual survey of vessel support blocks,
9. Underwater survey, and
10. Cathodic protection system (if applicable).

6.2.1 Track Foundation Inspect the exposed portions of the track foundation for signs of corrosion, deterioration, failure from overload, and other deficiencies.

6.2.2 Tracks and Rails A marine railway track can be constructed of steel, concrete, timber, or some combination of these materials.

6.2.2.1 Track Environmental Zones The marine railway track has three different environmental zones which can affect the materials in different ways, as shown in Figure 6-1. These zones are

1. *Underwater Zone*: The portion of track and foundation that remains submerged at all times.
2. *Splash and Tidal Zone*: The portion of track, foundation, and cradle that get wet from tidal variations and wave splash. This is usually the zone of heaviest deterioration.
3. *Above Splash Zone*: The portion of track, foundation, cradle, and hauling house that is above the heavy splash from waves.

6.2.2.2 Track Inspection Visually inspect the track and the exposed portions of the foundation and track for the following:

1. Ice damage: Inspect for damage by ice in the tidal zone. Damage may include bent or broken members, derailed roller frames, "jacking" of track up off the piles, or out-of-line or out-of-grade track.
2. Impact damage: Inspect for damage by floating debris in tidal zone. Check for damage by anchors and vessel impact in the tidal zone or the underwater zone. Damage may include bent or broken members, derailed roller frames, or out-of-line or out-of-grade track.
3. Mud or debris on or near track: Inspect for mud over the tracks. Mud covering the track beams is acceptable as long as it does not cover the rails. (Mud may actually protect the track beams from deterioration.) However, mud on the track can cause the rollers or wheels to ride up on the mud and derail, possibly causing a cradle derailment. If mud has built up greater than the level of the rails, the cradle will have to push through this mud when backing down. This will add considerable force to the backing chain or cable and underwater sheaves, possibly overloading them. Determine the mudline over the full width of the cradle to verify that no part of the cradle will contact the mud during cradle movement.
4. Gaps between pile and track beam: Gaps between pile and track beam indicate shims have fallen out, pile has sunk, or track has come up. Inspect track grade and shims inserted to suit. There shall be no gaps between pile and track, or the

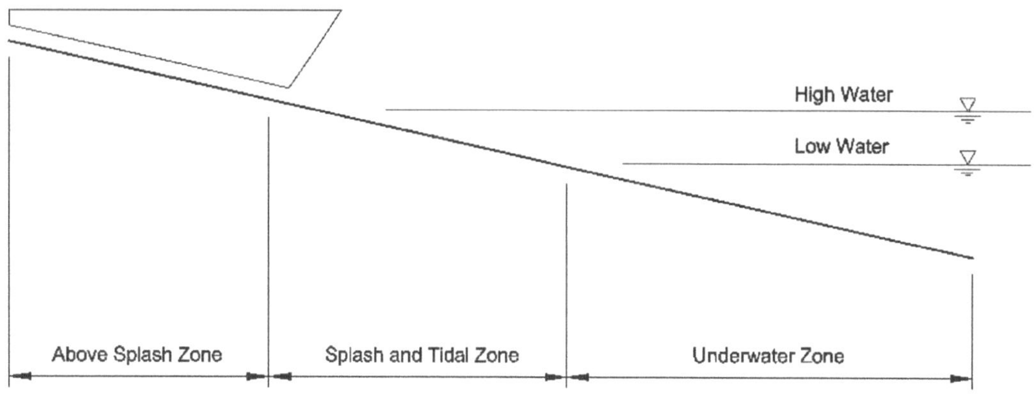

Figure 6-1. Typical marine railway deterioration zones.

track will deflect when the loaded cradle travels over this area.

5. Chain or cable slides: Inspect for worn, loose, or missing chain slides.

6.2.2.3 ***Rail Inspection*** Inspect the rail plate (or rails) for

1. Wear and thickness. Portions of rail plate which measured outside of the allowable wear and thickness tolerances shall be repaired or restored to allowable tolerances or replaced altogether.
2. Rail transitions. Inspect for smooth transitions between adjacent rail plates (or rails), with no sudden changes in rail plate width or thickness.

6.2.2.4 ***Concrete Track*** Inspect a concrete track for

1. Spalling: Some minor spalling is acceptable, but if spalling has progressed to undermine the rail plate or is exposing the rebar, the damage shall be repaired (Figure 6-2).
2. Major Cracking: Indicates possible overload, foundation failure, or impact damage.
3. Deteriorated Fastenings: Rails can pull away from the concrete or become loose, leading to a potential cause of derailment.

6.2.2.5 ***Steel Track*** A steel track deteriorates at varying rates over its length. Some areas of the track may look fine, whereas other areas have heavy corrosion. Inspect the entire length of track, above and below water. In general, the area of worst corrosion is usually the splash and tidal zone.

Inspect the condition of steel structural members. Determine the severity of corrosion and document the percentages of total area that are corroded. Document defects or damage such as bent or buckled frames, plates, or bulkheads; holes in plates; and cracks in plating, framing, or welds. Inspect a steel track for

1. Corroded steel;
2. Rolled or bent top flange: Overloading or off-center loading of the rail plate may cause the top flange of the track to roll or bend about the web of the beam (Figure 6-3); and
3. Missing rubber pads: Some steel tracks on steel piles have rubber pads installed between the track and pile bonnets.

6.2.2.6 ***Timber Track*** Inspect timber for rot, impact damage, signs of overload, and marine borers or other insects. Use a probe, core borer, and/or hammer to test wood. Determine the depth at which sound wood is found below punky surface.

Inspect the track for rail plate squeeze. The rail plate sometimes squeezes into the wood, reducing the height of the top of plate above the wood. The roller flanges can then contact the wood and wear grooves in the wood on either side of the rail plate (Figure 6-4). The roller flanges contacting the wood can create considerable additional friction that the hauling system must overcome. This additional force could overload the hauling system. Also, roller flanges may tend to break more often.

Inspect the track system for termites or other insects, usually found above the splash zone. Inspect the track system for marine

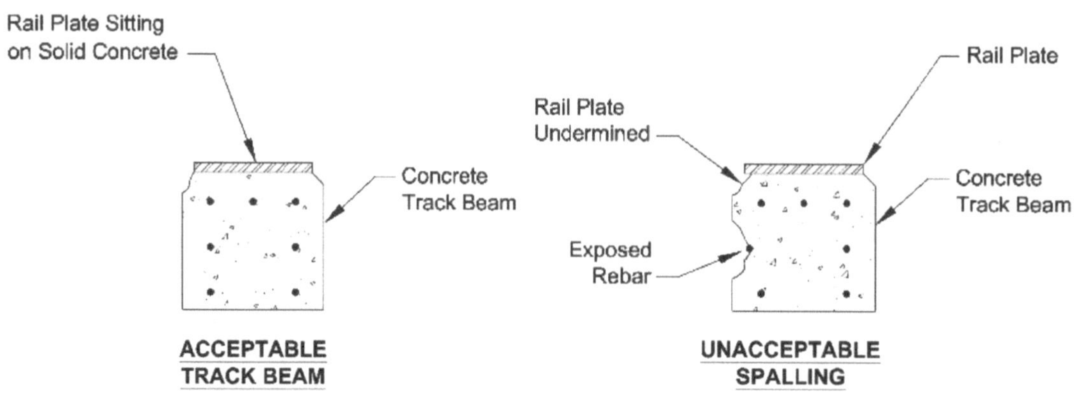

Figure 6-2. Spalling of track beam.

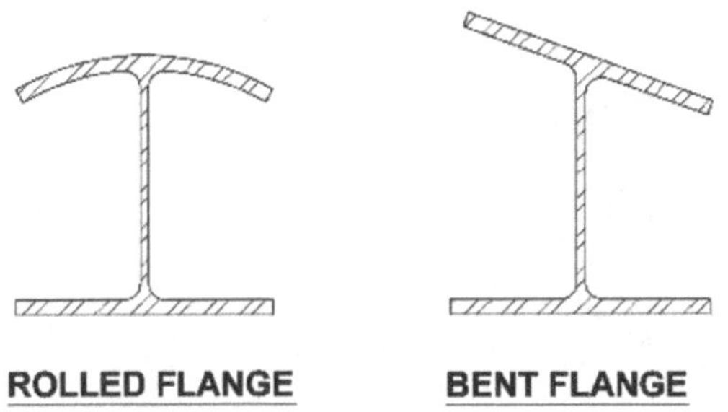

Figure 6-3. Deformed top flange.

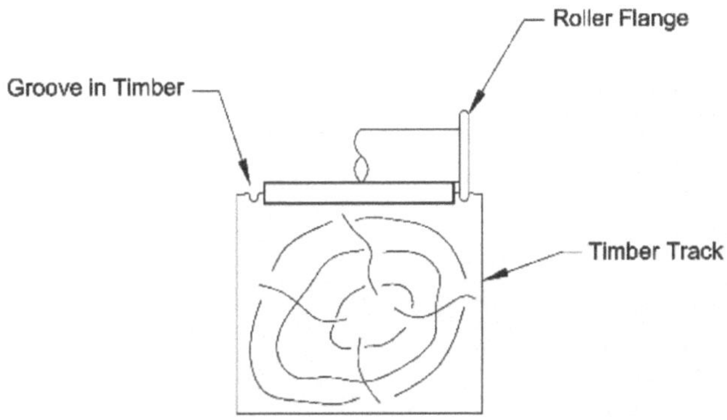

Figure 6-4. Rail plate squeeze.

borers, usually found in the tidal and underwater zones in salt water only. Two common types of marine borers are the wood gribble (*Limnoria lignorum*) and the shipworm (*Teredo navalis*). Marine borers are not always evident from the surface. Limnoria damage is readily evident with a visual inspection as loss of material occurs on the outside of the timber. *Teredo navalis* is a mollusk, although it looks like a worm. It bores a tiny hole into the wood when young, and as it grows it tunnels along the grain of the wood. Detection of *Teredos navalis* is difficult because most of the damage occurs inside the timber. A member may be riddled with holes but look fine from the outside (Figure 6-5).

Inspect the condition of track timber sheathing (if any). Sheathing may need to be removed in representative areas to check condition of tar coating, sheathing felt, and timber beneath. Inspect the track system for rotted timbers, usually found above the splash zone and in the splash and tidal zone.

6.2.3 Drydocking Cradle Inspect all structural elements of the cradle for corrosion; cracked, bent, or buckled members; or rotted or broken timbers. Inspect the cradle shoe plate, if so equipped, and check for wear and crowning in the same manner as the track rail plates.

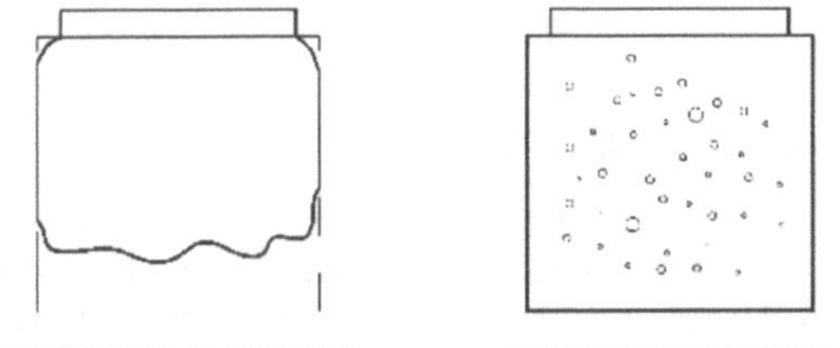

Figure 6-5. Marine borer damage.

6.2.4 Rollers and Wheels

6.2.4.1 Marine Railways Operated on System of Rollers For marine railways that operate on a system of rollers, inspect and document findings for the following

1. Roller Frames: Verify that the number and position of roller frames are adequate to allow the cradle to travel the full length of the track without overrunning the rollers or causing the rollers to run off the end of the track. Identify corrosion, wear, and bent or buckled members and verify that roller frames are properly connected together.
2. Rollers: Identify signs of wear, broken treads, or broken flanges. Measure a representative sample of roller treads to determine the amount of tread wear and to ensure rollers are within acceptable tolerance.
3. Roller pintles and bushings: Identify wear.

6.2.4.2 Marine Railways Operated on System of Wheels For marine railways that operate on a system of wheels, inspect and document findings for the following:

1. Wheels: Identify signs of wear, broken treads, or broken flanges. Measure a representative sample of wheel diameters to determine wheel wear.
2. Axles: Identify corrosion, wear, or cracks.

6.2.5 Chains or Wire Rope

6.2.5.1 Hauling and Backing Chain For marine railways that operate with a system of hauling chains and backing chains, inspect and document findings for the following

1. Chain links: Identify wear or cracks. Measure the chain thickness and ensure thickness is within acceptable tolerance. If outside of thickness tolerance, calculate the reduced chain capacity and revise certified lift capacity if required.
2. Connection between the backing chain and hauling chain.
3. Connection of the hauling chain to the drydocking cradle.
4. Drydocking cradle hauling chain equalizing sheaves (if applicable).
5. Drydocking cradle backing chain equalizing sheaves (if applicable).
6. Connection of the backing chain to the drydocking cradle.
7. Underwater sheave cases, sheaves, and pins. If excessive wear of the sheave pins is suspected, the sheave cases shall be brought to the surface and disassembled for closer inspection.
8. Verify that the chain system is properly tensioned (Figures 6-6 and 6-7).

6.2.5.2 Wire Rope For marine railways that operate with a wire rope hauling system, inspect and document findings for the following:

1. Wire rope;
2. Connection of the haul-in wire rope to the drydocking cradle;
3. Connection of the haul-out wire rope to the drydocking cradle; and
4. Underwater sheave cases, sheaves, and pins, if so equipped. If excessive wear of the sheaves or sheave pins is suspected, the sheave cases shall be brought to the surface and disassembled for closer inspection.

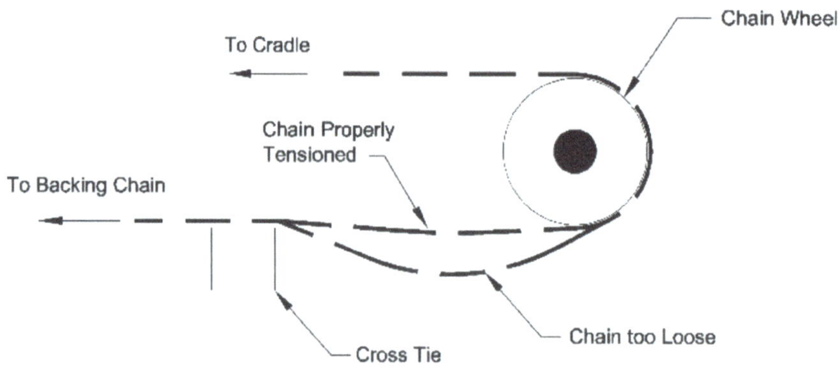

Figure 6-6. Hauling chain tension.

Figure 6-7. Loose hauling chain (right).
Source: Courtesy of Heger Dry Dock, Holliston, Massachusetts.

6.2.6 Hauling Machinery Observe the hauling machine in operation throughout an entire cycle of haul-out and haul-in. Provide documentation of findings.

6.2.7 Visual Survey of Drydocking Cradle Fenders and Vessel Moorings Inspect and document the condition of foundation plating around all cleats, bollards, winches, capstans, and other components, especially where the component fastens to the drydocking cradle which is, in general, an area highly susceptible to corrosion. Inspect the condition of timber fenders and attachments to wingwalls, especially at the interface, where wood tends to rot and steel tends to corrode.

6.2.8 Vessel Support Blocks Visually assess the condition of the vessel blocking system. Inspect the timber for excessive crushing, warping, cracking, checking, rotting, or wear from fasteners. Inspect for loss of contact at edges. In general, minor cracking is acceptable unless bearing area has been lost. If the blocks have concrete bases, inspect the concrete for cracking, spalling, and exposed rebar. If the blocks have steel bases, inspect the steel for loss of metal thickness by corrosion, cracks, cracked welds, and other defects. Inspect the condition of fastenings. All timber shall be adequately secured to prevent floating when the dock is submerged.

If the marine railway is equipped with hauling side blocks, inspect the hauling block hardware. The hauling block hardware includes chains, sheaves, cranks, slides, and pawls.

6.2.9 Underwater Survey Conduct an underwater survey of the submerged portions of the exposed foundation; track, chain, or wire rope system; and cradle.

6.2.10 Cathodic Protection System Inspect the cathodic protection system (if applicable).

6.3 OPERATIONAL TESTS

In addition to the material condition survey, dry dock operational tests are required to establish the existing condition and function of the dry dock. In general, the operational tests shall include (1) dry dock operation test and (2) verification of instrumentation and controls.

6.3.1 Dry Dock Operation Test The surveyor shall witness one complete cycling of the marine railway to include the full length of track to be certified. The operational test shall include the drydocking or undocking of a vessel. The size and displacement weight of the vessel in relation to the operational limitations and capacity of the dry dock shall be at the discretion of the certifying authority. The marine railway's mechanical and electrical components and associated control systems shall be operated by the shipyard and witnessed by the surveyor. The operational test shall include the following:

1. Hauling machinery,
2. Primary power source,
3. Controls,
4. Ammeters,
5. Limit switches,
6. Alarms,
7. Capstans,
8. Winches,
9. Vessel centering system (if so equipped),
10. Alarms,
11. Lighting,
12. Fire protection system, and
13. Drydocking cradle locks.

6.3.2 Verification of Instrumentation and Controls Verify that all instrumentation and controls are in good working order. If, during the operational test, ammeter gauges, position indicators, or other instrumentation do not reflect the actual condition, the errant gauges shall be recalibrated and checked by comparing the physical measurement to the indication on the gauge.

6.4 DRY DOCK CALCULATIONS

Establish, verify, or adjust the capacity, stability, and operational limitations of the marine railway in its present condition. The dry dock evaluation shall include the following:

1. Capacity of foundation,
2. Capacity of track,
3. Structural strength of drydocking cradle,
4. Stability of drydocking cradle,
5. Capacity of drydocking cradle locking pin,
6. Capacity of hauling machine,
7. Capacity of hauling chain (or wire rope),
8. Block loading calculations,
9. Line, grade, and gauge tolerances, and
10. Roller tolerances.

6.4.1 Capacity of Foundation Determine the maximum allowable wheel or roller loads that the foundation is capable of supporting. This shall be checked against the maximum calculated loads induced on the foundation when docking a capacity vessel.

The loading on the foundation varies throughout the length of the track. At the end of the track, only the submerged weight of the cradle needs to be supported. As the cradle moves inshore and comes in contact with the vessel, the vessel's weight is gradually transferred onto the cradle. At the point where the ship's keel breaks the water, its full weight is on the cradle and the remaining portion of the track and foundation shall be analyzed for this condition. The foundation under the cradle in the full-up condition shall also be investigated for additional loads induced by extreme winds or earthquakes with a capacity vessel aboard (Figure 6-8).

6.4.2 Capacity of Track Determine the maximum allowable wheel or roller loads that the track is capable of supporting. This shall be checked against the maximum calculated loads induced on the track when drydocking a capacity vessel.

6.4.3 Structural Strength of Drydocking Cradle Verify the ability of the cradle to support the maximum rated loads by structural analyses of

1. Cradle beams,
2. Cradle columns,
3. Bottom chords, and
4. Wheels or rollers.

An example of a marine railway drydocking cradle configuration is shown in Figure 6-9.

6.4.4 Stability of Drydocking Cradle Analyze the ability of the drydocking cradle to resist overturning in current, high winds, or earthquake loading and document findings.

6.4.5 Capacity of Drydocking Cradle Locking Pin Determine the capacity of the drydocking cradle locking pin.

6.4.6 Capacity of Hauling Machinery Determine the maximum allowable line pull on the hauling machinery.

6.4.7 Capacity of Hauling Chain (or Wire Rope) For dry docks designed with a chain hauling system, the allowable line pull on the chain shall be calculated using a minimum cross-sectional chain link area as defined by the surveyor. This is the chain link area that is required to haul a capacity vessel. The surveyor shall recommend the frequency of a maintenance task to inspect the chain to ensure that the chain maintains adequate section.

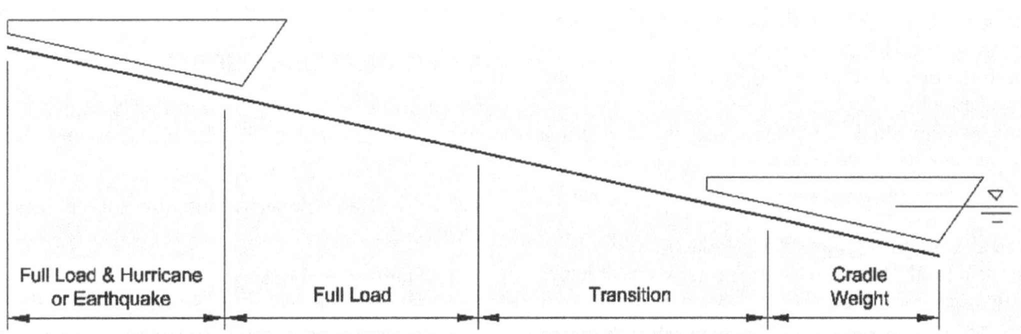

Figure 6-8. Typical zones of foundation and track loads.

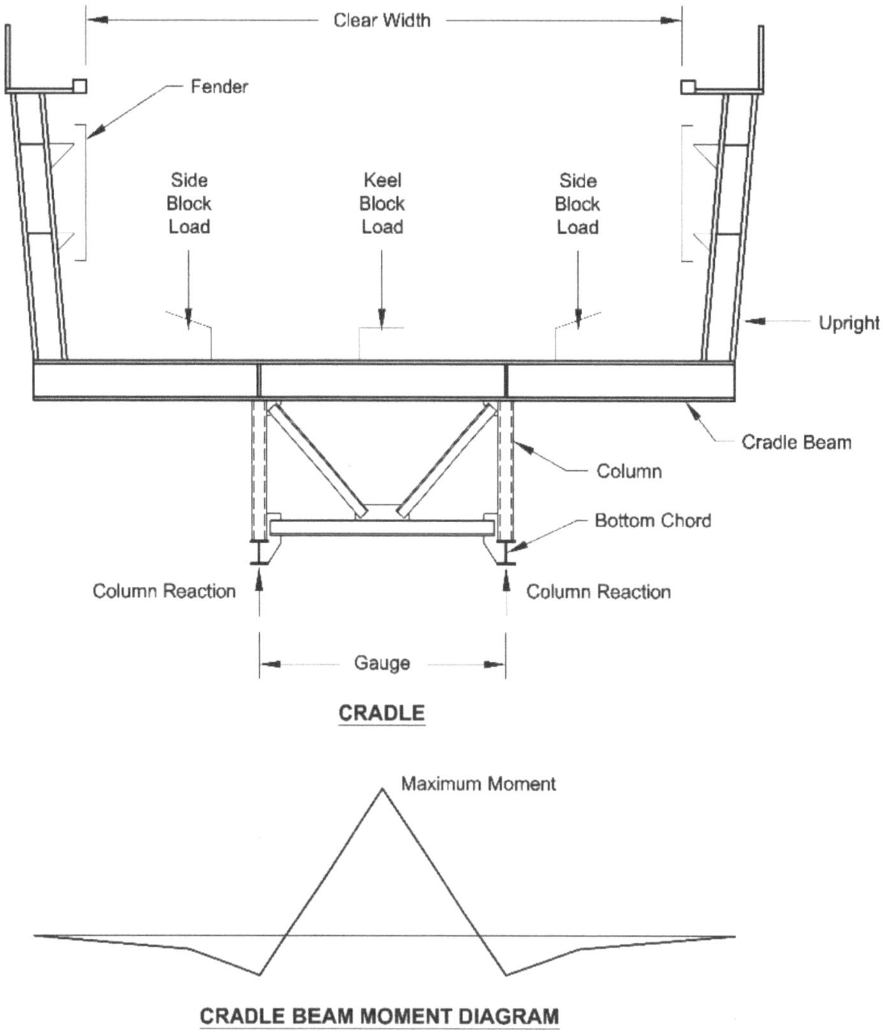

Figure 6-9. Marine railway drydocking cradle loading.

For dry docks designed with a wire rope hauling system, the allowable line pull on the wire rope shall be determined by the wire rope manufacturer's specified safe working load. Deteriorated wire rope shall be inspected and replaced as required.

6.4.8 Block Loading Analysis Perform calculations to determine that the vessel support blocks are stable and structurally adequate to support the marine railway's certified rated load per block. Calculations shall include

1. Allowable bearing pressure on timber times bearing area,
2. Structural strength of block base unit,
3. Local structural strength of marine railway structure directly supporting blocks,
4. Rated load per keel block. The rated load per keel block shall be the lesser of
   (a) Allowable bearing pressure on timber times bearing area, or
   (b) Structural strength of block base unit, or
   (c) Local structural strength of marine railway structure directly supporting blocks,
5. Rated load per side block. The rated load per side block shall be the lesser of
   (a) Allowable bearing pressure on timber times bearing area, or
   (b) Structural strength of block base unit, or
   (c) Local structural strength of marine railway structure directly supporting blocks.

6.4.9 Line, Grade, and Gauge Tolerances Determine the line, grade, and gauge tolerance of the railway.

6.4.10 Roller Tolerances Determine the roller tolerances.

6.5 CONDITION ASSESSMENT REPORT

In addition to the requirements of Chapter 3 of this standard, include in the written Condition Assessment Report the following items.

6.5.1 Written Descriptions Provide a description of the following

1. Hauling machinery,
2. Hauling chain (or wire rope),
3. Backing chain (or wire rope),
4. Method of monitoring hauling chain load,

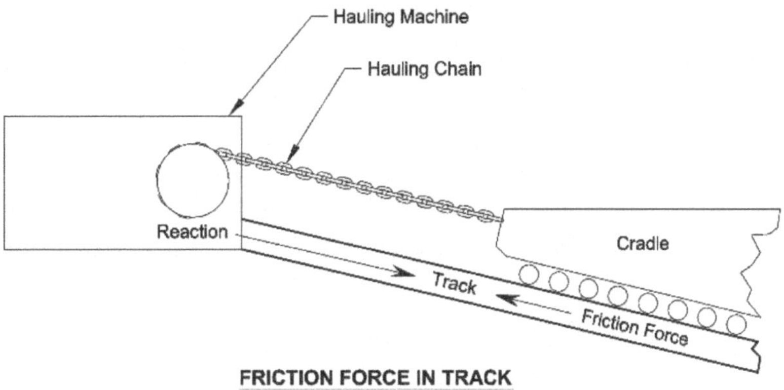

Figure 6-10. Marine railway forces.

5. Results of the line, grade, and gauge survey, and
6. Tolerances of the wheels in their existing condition.

6.5.2 Calculations Include and provide a written summary of the calculations required by Chapter 6 of this standard.

6.5.3 Overall Lift Capacity State the overall lift capacity of the marine railway. The overall lift capacity shall be based on the lesser of

1. Capacity of the hauling system machinery or chain (or wire rope), or
2. Rated load per foot capacity times the available keel blocking length, or
3. Maximum weight ship owing to cradle stability requirements.

When determining the overall capacity of the marine railway, it is important to establish the maximum coefficient of friction for the wheel or roller system (Figure 6-10). The coefficient of friction may vary, depending on the condition of the wheels or rollers, condition of the track, alignment of the track, or mud and debris on track.

6.5.4 Rated Load per Foot Capacity State the rated load per foot capacity of the marine railway. The rated load per foot capacity shall be the lesser of

1. Capacity of the drydocking cradle structure to support maximum rated line load,
2. Capacity of track structure to support maximum rated line load,
3. Capacity of foundation to support maximum rated line load,
4. Keel block capacity, or
5. Local structural strength of marine railway structure directly supporting blocks.

6.5.5 Allowable Deck Loading State the maximum allowable cradle deck loading. State the maximum allowable cradle deck vehicle loadings if the deck is to be used for vehicles such as material handling equipment or aerial manlifts.

CHAPTER 7
CONDITION ASSESSMENT REQUIREMENTS FOR VERTICAL LIFTS

7.1 DOCUMENT REVIEW

Documents governing the original design, capacity, construction, rehabilitation, alteration, and repair of the drydocking facility shall be reviewed. The design data required will vary depending on the type of vertical lift being evaluated. In general, the document review shall include

1. General arrangement of the vertical lift;
2. Environmental data, including seismic, water, and wind data;
3. Geotechnical data;
4. Design calculations;
5. Structural drawings and specifications;
6. Mechanical drawings and specifications;
7. Electrical drawings and specifications;
8. Instrumentation, control, and monitoring systems;
9. Operating manual/procedures;
10. Hydrographic data; and
11. Previous inspection and survey reports.

7.2 MATERIAL CONDITION SURVEY

A material condition survey is required to establish the existing condition of the drydocking facility. The condition of the finger piers, lifting platform structural steel, transfer rails of roadway, decking, and blocking (steel and/or timber) shall be assessed in accordance with industry standards and documented in the certification inspection checklists. Because most vertical lifts incorporate an extensive vessel transfer system that is not covered by this standard, only the transfer rails or rubber-tired transfer vehicle support paths that are part of the vertical lift platform shall be assessed. In general, the material condition survey for vertical lifts shall include

1. Visual survey of the lifting equipment support piers and piles;
2. Visual survey of vertical lift platform;
3. Inspection of platform lifting equipment (e.g., hoists);
4. Inspection of platform lifting equipment (wire ropes, chains, or other mechanisms);
5. Inspection and verification of vertical lift primary braking systems;
6. Inspection and verification of vertical lift control and protection equipment, including operational dry dock test and visual survey of all electric power, control, and signal wiring and conduits between the vertical lift motor control center, operator controls, and the lifting equipment and limit switches;
7. Visual survey of vertical lift fenders, vessel moorings on the support piers (bollards), and vessel haul-in system (capstans and fairleads); and
8. Visual survey of vessel blocking support system.

7.2.1 Support Piers and Piles Inspect the support piers and piles for signs of corrosion, deterioration, failure from overload, and other deficiencies. Excavation of the below-grade foundation is usually not required unless foundation failure is suspected. Inspection is to include deterioration that may create falling debris that could affect the platform lifting equipment, such as wire ropes.

7.2.1.1 Underwater Survey Inspect the underwater portions of the pier piles and bulkhead.

7.2.1.2 Cathodic Protection System Inspect the cathodic protection system (if applicable). Verify that all electric power to any impressed current (active) cathodic protection system will be automatically or manually turned off when the vertical lift platform is immersed to protect the wire ropes and all equipment that can become a grounding path (if applicable). Verify the grounding paths are correct.

7.2.2 Vertical Lift Platform Inspect all structural elements of the vertical lift platform for corrosion and cracked, bent, or buckled members. Inspect the platform decking for suitability of design purpose and potential docking hazards.

7.2.2.1 Wire Rope For vertical lifts that operate with a wire rope lifting system, inspect and document findings for the following:

1. Wire rope condition: Rope rejection for wear, corrosion, and damage to be determined through applying a defined rejection criteria from either the original equipment manufacturer, if available, or through the minimum criteria as defined by a classification society code that specifically addresses the use of wire rope for a vertical lift in a marine environment. More stringent rejection criteria can be defined by the certification agency.
2. Effectiveness of wire rope lubricant and frequency of application to maintain wire ropes through next survey.
3. Age and operational in-service hours for each wire rope. Ropes in service for more than 10 years or 2,000 h to be replaced.

End connections of the wire rope to the lifting equipment, if applicable.

Inspection of the equipment within the wire rope system running path, which includes sheaves and the hoist drum.

7.2.2.2 Chains and Other Lifting Systems For vertical lifts that operate with chains or other lifting systems, inspect and document findings for the following:

1. Chain or other lifting system condition: Chain or other equipment rejection for wear, corrosion, and damage to be determined through a defined rejection criteria either by the original equipment manufacturer, if available, or as defined under a recognized standard for inspection of the type of lifting equipment being used, including documentation. More stringent rejection criteria can be defined by certification agency.
2. Effectiveness of lubrication or other protection to maintain chain or other lifting equipment through next survey.
3. Connection of the chain or other lifting system to the platform and/or lifting equipment.
4. Inspection of the equipment within the chain or other lifting equipment system running path.

7.2.3 Vertical Lift Drive and Primary Brake System Visual inspection and verification of the vertical lift drive system (such as electrical or hydraulic motors). Visual inspection and verification of primary brake system (such as motor brakes or drum brakes). Include documentation of any secondary brake or support systems on the vertical lift.

7.2.4 Vertical Lift Control Systems Perform a visual inspection of the utility runs between the motor control center and the vertical lifting equipment including electrical power cables, control and signal wires, conduit, compressed air system, and/or hydraulic system.

7.2.5 Lifting Hoists The lifting hoist shall be observed in operation throughout an entire cycle of lowering and raising the platform. Provide documentation of findings.

7.2.6 Vertical Lift Fenders and Vessel Moorings Inspect and document the condition of foundation plating around all cleats, bollards, winches, capstans, and other components. Inspect the condition of timber fenders and attachments to piers, especially at the interface, where wood tends to rot and steel tends to corrode.

7.2.7 Vessel Support Blocks Visually assess the condition of the vessel blocking system. Inspect the timber for excessive crushing, warping, cracking, checking, rotting, or wear from fasteners. Inspect for loss of contact at edges. In general, minor cracking is acceptable unless bearing area has been lost. If the blocks have concrete bases, inspect the concrete for cracking, spalling, and exposed rebar. If the blocks have steel bases, inspect the steel for loss of metal thickness by corrosion, cracks, cracked welds, and other defects. Inspect the condition of fastenings. All timber shall be adequately secured to prevent floating when the platform is submerged.

Inspect all load path supports for the blocking system.

If the vertical lift is equipped with hauling side blocks, inspect the hauling block hardware. The hauling block hardware includes chains, sheaves, cranks, slides, and pawls.

7.3 OPERATIONAL TESTS

In addition to the material condition survey, dry dock operational tests are required to establish the existing condition and function of the dry dock. In general, the operational tests shall include

1. Dry dock operation test, and
2. Verification of instrumentation and controls.

7.3.1 Dry Dock Operation Test The surveyor shall witness one complete cycling of the vertical lift. The operational test shall include the drydocking or undocking of a vessel. The size and displacement weight of the vessel in relation to the operational limitations and capacity of the dry dock shall be at the discretion of the certifying authority. The vertical lift's mechanical and electrical components and associated control systems shall be operated by the shipyard and witnessed by the surveyor. The operational test shall include the following:

1. Lifting hoist. The synchronization of the hoists is a key characteristic to observe because this is essential for the safe operation of the vertical lift,
2. Load monitoring system,
3. Limit switches,
4. Alarms,
5. Emergency backup power,
6. Vessel handling system,
7. Lighting,
8. Fire protection system, and
9. If the vertical lift allows for direct side transfer to and from the platform, the platform locking system shall be demonstrated.

7.3.1.1 Platform Lifting Equipment (Such as Hoists) The lifting hoists shall be observed in operation throughout an entire cycle of lowering and raising the platform. Provide documentation of findings. Inspect gear-train system using suitable means for the type of equipment.

7.3.1.2 Wire Rope or Chain Lifting Systems The lifting hoists shall be observed in operation throughout an entire cycle of lowering and raising the platform. Provide documentation of findings. Inspect gear-train system using suitable means for the type of equipment.

7.3.2 Control and Protection Tests Inspect and document findings for the following:

1. Ensure that equipment for maintaining the platform level through the drive system for keeping the lifting equipment line pull speed synchronized is working correctly;
2. Protection for lifting equipment overloading and slack rope or chain conditions is operating correctly;
3. Protection for lifting equipment upper and lower limit protection is operating correctly;
4. Verification of all other original primary protection systems;
5. Emergency backup power (if available);
6. Lighting;
7. Fire protection system; and
8. If the vertical lift allows for direct side transfer to and from the platform, the platform locking system shall be demonstrated.

7.4 DRY DOCK CALCULATIONS

Establish, verify, or adjust the capacity and operational limitations of the vertical lift in its present condition. The dry dock evaluation shall include the following:

1. Capacity of foundation,
2. Capacity of support piers,
3. Structural strength of platform,
4. Capacity of lifting hoist,
5. Capacity and operational requirements of wire ropes, and
6. Block loading calculations.

Figure 7-1 shows the arrangement of a typical vertical lift.

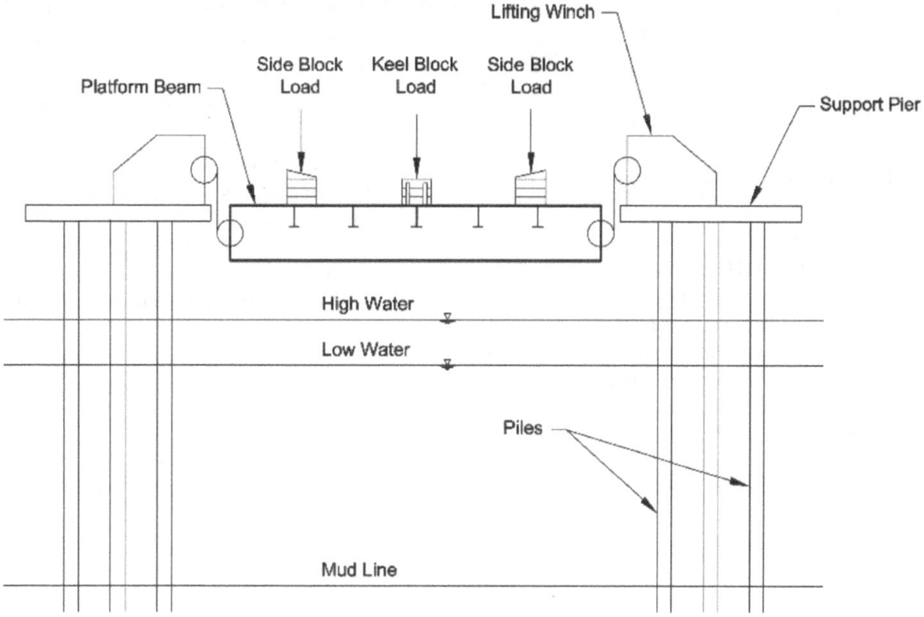

Figure 7-1. Vertical lift cross section.

7.4.1.1 Capacity of Foundation Determine the maximum allowable hoist loads that the foundation is capable of supporting. This shall be checked against the maximum calculated loads induced on the foundation when docking a capacity vessel.

7.4.1.2 Capacity of Support Piers Determine the maximum allowable hoist loads that the support piers are capable of supporting. This shall be checked against the maximum calculated loads induced on the foundation when docking a capacity vessel.

7.4.1.3 Structural Strength of Platform Verify by structural analysis the ability of the platform to support the maximum rated loads.

7.4.1.4 Capacity of Lifting Hoist Determine the maximum allowable line pull on the lifting hoists.

7.4.1.5 Capacity and Operational Requirements of Wire Ropes Determine the capacity of the wire rope lifting system and the maximum allowable line pull on the wire rope.

7.4.1.6 Block Loading Analysis Perform calculations to determine that the vessel support blocks are stable and structurally adequate to support the vertical lift's certified rated load per block. Calculations shall include

1. Allowable bearing pressure on timber times bearing area.
2. Structural strength of block base unit.
3. Local structural strength of the vertical lift's platform structure directly supporting blocks.
4. Maximum allowable block load based on the maximum allowable design load (MDL) of the lifting platform per unit length. The MDL is calculated by subtracting the total weight of the platform between two main transverse lifting beams supported at the ends by the lifting hoists from the capacity of two hoists and dividing by the hoist spacing.
5. State the rated load per keel block. The rated load per keel block shall be the lesser of
   (a) Allowable bearing pressure on timber times bearing area,
   (b) Structural strength of block base unit,
   (c) Local structural strength of vertical lift structure directly supporting blocks, or
   (d) Maximum allowable block load based on the MDL of the lifting platform per unit length. The MDL is calculated by subtracting the total weight of the platform between two main transverse lifting beams supported at the ends by the lifting hoists from the capacity of two hoists and dividing by the hoist spacing.
6. State the rated load per keel block. The rated load per keel block shall be the lesser of
   (a) Allowable bearing pressure on timber times bearing area,
   (b) Structural strength of block base unit,
   (c) Local structural strength of the vertical lift's platform structure directly supporting blocks, or
   (d) One-half of the adjacent keel block load.

7.5 CONDITION ASSESSMENT REPORT

In addition to the requirements of Chapter 3 of this standard, include in the written report the items listed in the following sections.

7.5.1 Written Descriptions Provide a description of the following:

1. Platform construction,
2. Platform lifting equipment,
3. Lifting wire rope, chains, or other equipment,
4. Platform lifting equipment drive and primary brake system,
5. Control system and primary protection equipment, and
6. Vessel blocking support system.

7.5.2 Maximum Lift Beam Capacity State the maximum allowable live load per lift beam and the allowable distribution of the load on the beam. The maximum allowable live load per lift beam shall be based on the lesser of

1. Strength of the lift beam,
2. Capacity of the wire rope and sheave system,
3. Strength of the connection of the lower sheave system to the ends of the lift beam,
4. Wire rope attachment to the lift beam (if applicable),
5. Capacity of the hoist, or
6. Capacity of the pier and foundation to support the hoist loads.

7.5.3 Load per Foot Capacity (MDL) State the maximum design load per foot capacity of the vertical lift. The MDL is calculated by subtracting the total weight of the platform between two main transverse lifting beams that are supported at the ends by the lifting hoists, from the capacity of the two hoists and dividing by the hoist spacing. The maximum load per foot capacity shall be the lesser of the

1. Maximum allowable live load per lift beam divided by the hoist spacing, or
2. Keel block or transfer car capacity.

7.5.4 Vessel Support Blocks Load Limits State the rated load per keel block. The rated load per keel block shall be the lesser of

1. Allowable bearing pressure on timber times bearing area, or
2. Structural strength of block base unit, or
3. Local structural strength of dock structure directly supporting blocks.
4. Rated load per side block. The rated load per side block shall be the lesser of
5. Allowable bearing pressure on timber times bearing area, or
6. Structural strength of block base unit, or
7. Local structural strength of dock structure directly supporting blocks.

State the maximum allowable platform deck loading. State the maximum allowable platform deck vehicle loadings.

CHAPTER 8
CERTIFICATION INSPECTION REQUIREMENTS

8.1 PERIODICITY OF CERTIFICATION INSPECTIONS

The periodicity of dry dock certification inspections shall be at the discretion of the certifying agency but shall not exceed 3 years if a facility is less than 10 years old, and no more than 2 years if the facility is more than 10 years old.

8.2 DOCUMENT REVIEW

The following data shall be made available to the certifying agency for review during certification inspections:

1. As-built facility drawings,
2. Major modification and/or repair drawings,
3. Condition Assessment Report including the original dry dock certification certificate,
4. Previous certification inspection reports,
5. Capacity and stability calculations (as applicable),
6. Engineering justification for any deferred inspection items, and
7. Preventive maintenance records in electronic or printed form.

8.3 CERTIFICATION INSPECTION SCOPE

8.3.1 Review of Existing Data The certifying agency shall review the facility drawings, the Condition Assessment Report, and previous certification inspection reports.

8.3.2 Document and Calculation Review The certifying agency shall review drawings, the Condition Assessment Report, previous certification inspection reports, and calculations prior to conducting certification inspections to identify areas warranting closer attention during the material condition survey. The certifying agency shall verify or adjust the capacity, stability (as applicable), and operational limitations of the dry dock based on its present (existing) condition. If the dry dock's structure and/or equipment has deteriorated to the point at which the original capacity, stability and operational limitations are questionable; then a new dry dock condition assessment is required.

The certifying agency shall also review maintenance records to verify that proper maintenance of the facility is being performed regularly and determine what maintenance has been recently performed.

If necessary, the certifying agency shall also use this information to recommend changes to shipyard maintenance and/or operational procedures.

8.3.3 Material Condition Survey A material condition survey is required to establish the existing condition of the drydocking facility. The certifying agency's surveyor shall ascertain the existing condition of all critical components of the facility and verify that the certified rated capacity, stability limitations (as applicable), and operational limitations are still valid. The certification inspection shall identify material defects and evaluate the effectiveness of the shipyard's preventive maintenance and deficiency correction procedures. All findings and deficiencies shall be documented on the certification inspection checklists and included in the certification inspection report. The certification inspection checklists developed during the condition assessment or from past certification inspections shall be used as a baseline and modified by the certifying agency as required.

In areas where section loss is suspected, such as areas of failed or missing protective coating, areas of high corrosion, or areas where the as-built member thicknesses are in question, the use of UT measurements, calipers, or other devices to measure the steel wastage shall be utilized to establish the remaining steel thickness. The certifying agency responsible for the certification inspection shall establish the frequency and location of the measurements, which will vary depending on the condition of the steel.

Cathodic protection systems shall be inspected, where applicable, by qualified personnel.

8.3.4 Operational Inspections Mechanical and electrical systems, equipment, and components shall be inspected and operationally tested in the presence of the certifying agency. Routine maintenance of the systems counts as an acceptable inspection if the maintenance activity is documented and can be proven to the inspector that the system(s) will perform their intended functions.

A complete operational inspection of the facility and its mechanical and electrical systems shall be conducted, and the condition compared to the Condition Assessment Report as well as recent certification inspection reports. Inspections may be performed in coordination with scheduled dry dock operations.

The operational inspection shall determine the overall capability of the mechanical and electrical systems as they relate to the proper functioning of the facility, by observing the operation of all the facilities equipment. Observation of equipment includes visual and audible assessment. It includes, but is not limited to, motors or engines checked for smooth running without excessive vibration or unusual noise; valve guides inspected to determine the operational capacity, limit settings, and ride height; tank and draft level indicators verified; and deflection monitoring systems verified.

The data compiled from the visual survey shall be used to develop monitoring plans, recommend repair or replacement, or recommend a larger engineering study be conducted to determine the full extent of a detected issue.

8.3.5 Certification Inspection Checklists The material condition survey shall be documented in the certification inspection checklists. Sample certification inspection checklists are provided in Appendix A. The certification inspection checklists used shall be specific to the type of drydocking facility.

8.4 SPECIFIC CERTIFICATION INSPECTION REQUIREMENTS

8.4.1 Floating Dry Docks The buoyant capacity shall be verified at least once every 3 years. The buoyant capacity of the dry dock can change over time. The accumulation of mud, sandblast grit, addition of equipment, or the lack of ability of the pumping system to dewater the ballast tanks as low as originally designed can reduce the buoyant lift capacity. Verification of the buoyant capacity shall be accomplished by a lightweight survey and verification of the floating dry dock's weight in the capacity calculations.

8.4.2 Graving Docks At this time, there are no specific certification inspection requirements for graving docks.

8.4.3 Marine Railways At this time, there are no specific certification inspection requirements for marine railways.

8.4.4 Vertical Lifts During the certification inspection, certain items shall be verified as being inspected and maintained following the manufacturer's and classifying society's criteria. The certification inspection shall include the following actions/information:

1. Record the hours of operation since the previous inspection.
2. Review the vertical lift maintenance records and operating procedures.
3. Visually check the condition of each hoist foundation, including pile caps and piles above the water line.
4. Visually inspect condition of utility runs to each hoist (power, control, load monitoring system cables, and, if applicable, air piping).
5. Inspect the vertical lift platform steel, decking, and lower sheave assemblies.
6. Using data from the load cell system, calculate the wire rope sheave efficiency.
7. Each year, the wire ropes (or chains) shall be maintained and inspected so as not to lead to deterioration beyond the rejection criteria.
8. Visually inspect all hoist frames, covers, couplings, bearing blocks, and mountings.
9. Examine all hoist brakes, test the torque, and adjust or take remedial action, as necessary.
10. Test the insulation resistance of all field cables and motors to ground.
11. Visually check all switches (including hoist limit switches), motor starters, and relays for condition, proper operation, and adjust relay settings as required.
12. Check the entire electrical system (power and control) and verify proper operation in all modes of operation.
13. Verify the proper operation of the load monitoring system, if so equipped, and adjust as required.
14. Visually inspect the general condition of the transfer rails and rail securing clips on the platform (if applicable).

8.5 INSPECTION ITEM DEFERRAL

The shipyard may elect to defer some items in the certification inspection to accommodate dry dock operation or other significant conflicts that prevent accomplishment on schedule. Deferred certification inspection items shall be documented in the certification inspection checklists and shall be accomplished at the first opportunity. Certification inspections shall not be deferred more than 6 months from the inspection item's original interval without engineering justification and concurrence from the certifying agency for the extended deferral.

8.6 CORRECTION OF DEFICIENCIES

Any deficiencies identified during the certification inspections which affect the capacity or the safety of the facility shall be listed on the Dry Dock Capacity Certificate. All deficiencies rated MARGINAL or UNSATISFACTORY shall be added to the shipyard's maintenance program and tracked until the deficiency is corrected. The results of the certification inspection shall be evaluated to determine whether there is a need to repeat the condition assessment.

8.7 DRY DOCK CERTIFICATION REPORT

On completion of the dry dock certification inspection, the certifying agency shall prepare a written report summarizing the findings of the certification inspection. At a minimum, the certification inspection report shall include

1. Scope of the certification inspections performed;
2. Experience and qualifications of the certifying agency key personnel and key personnel who conducted the certification inspections;
3. List of all drawings, reports, calculations, and other data reviewed;
4. Summary of the material condition survey;
5. All findings in the certification inspection checklists;
6. Dates of the material condition survey;
7. List of items not inspected and reasons for lack of inspection;
8. List of equipment which was not operated and reasons for lack of operation;
9. List of all MARGINAL conditions along with a disposition;
10. List of all UNSATISFACTORY conditions along with a disposition;
11. Pertinent photographs and/or video recordings;
12. Dry Dock Capacity Certificate;
13. Recommendations for adjustments to maintenance procedures (if required);
14. Recommendations for adjustments to operating procedures (if required);
15. Recommendation regarding the frequency and scope of future certification inspection; and surveys based on the condition of the dry dock and the operating environment.

CHAPTER 9
DRY DOCK MAINTENANCE

9.1 MAINTENANCE PROGRAM

The shipyard shall have a documented maintenance program to ensure reliable operations of all critical equipment which serves the dry dock, and that the drydocking facility remains in a condition to safely operate at the certified rated capacity. The maintenance program shall consist of scheduled preventive/predictive maintenance as well as a process for tracking the correction of deficiencies.

Records of completed maintenance, both scheduled and emergent, shall be made available to the certifying agency for audit.

If a deficiency is observed during maintenance activities that results in an UNSATISFACTORY condition to structure or equipment critical to dry dock capacity, stability (as applicable), or operation, the certifying agent shall be notified within 48 hours.

Appendix B contains examples of maintenance tasks for the various types of dry docks which may be used as a guide. The detailed maintenance tasks should be specific to the drydocking facility and based on original equipment manufacturer's recommendations, if available.

CHAPTER 10
DRY DOCK OPERATIONS—GENERAL REQUIREMENTS

10.1 DRY DOCK OPERATIONS PERSONNEL

10.1.1 Dry Dock Operations Personnel Qualifications All dry dock operations personnel shall be qualified through training and experience. Individual qualifications shall be documented and retained on file as a matter of individual personnel records. The Dockmaster and/or Shipyard Management shall approve dry dock personnel qualifications, making the certifications a matter of record.

Dry dock operations personnel shall be actively involved in the position for which they are stationed, in at least three drydocking and three undocking operations with the type of dry dock for which they are qualified.

10.1.2 Dockmaster Shipyard Management shall designate a Dockmaster for each drydocking evolution.

10.1.2.1 Dockmaster Responsibilities The Dockmaster is responsible for the safety of the vessel and the safe operation of the dry dock during a drydocking evolution. During drydocking operations, the Dockmaster is responsible for the safety of the vessel, beginning when the first part of the vessel crosses the sill (or opening) of the dry dock. The responsibility for vessel safety is maintained throughout drydocking operations and the lay period. During undocking operations, the Dockmaster is responsible for the safety of the vessel until the last part of the vessel crosses the sill (or opening) of the dry dock. The transfer of responsibility for the vessel shall be communicated and acknowledged by a verbal and/or radio call between the Dockmaster and the Pilot or Master.

10.1.2.2 Qualifications of the Dockmaster The Dockmaster shall be qualified through training and experience to direct all drydocking evolutions in a safe and reliable manner. The Dockmaster shall be qualified for each type of drydocking facility using the criteria outline in this standard and shall be approved by Shipyard Management, making the Dockmaster qualifications a matter of record.

10.1.2.3 Dockmaster Training Shipyard Management shall document that the Dockmaster has received formal training in the applicable dry dock procedures and drydocking calculations used in this standard.

10.1.2.4 Dockmaster Experience To be qualified as a Dockmaster, the individual shall first serve under a qualified Dockmaster as a Dockmaster-in-training, being actively involved in a total 10 drydocking and 10 undocking operations. Five drydocking and five undocking operations shall have been completed on the type of dry dock for which the individual is to be qualified to serve as a Dockmaster.

The Dockmaster shall keep a log documenting their experience.

10.1.2.5 Retaining Dockmaster Qualification To remain qualified, the Dockmaster shall participate in at least one drydocking evolution every year with the type of dry dock for which the Dockmaster is qualified. This qualification requires the review of drydocking plans and checking corresponding block buildup requirements, as well as active involvement in the actual drydocking and undocking operations.

10.2 OPERATING PROCEDURES

The shipyard shall develop, maintain, and implement a series of operating procedures for the safe use of each drydocking facility. Prerequisite procedures and checks are required prior to commencing drydocking, vessel transfer (if applicable), and undocking operations. All dry dock operating procedures shall be established in step-by-step detail. The documented procedures are a matter of record and shall be strictly followed.

Sequential checklists shall be developed for all dry dock operating procedures. Appendix C is provided as a guideline to develop site specific checklists for each facility.

Methods of communication shall be included in the operating procedures. These methods may include two-way radios, push-to-talk phones, cell phones, and/or hand signals. Potential for radio frequency interference with the drydocking system controls shall be determined before allowing communication equipment near sensitive equipment.

Emergency procedures shall be developed for the rapid response to unexpected events, including but not limited to, bad weather, fire, uncontrolled flooding, operational accidents, equipment failures, and personnel injuries.

10.3 DRY DOCK OPERATION MANNING PLAN

A manning plan to support dry dock operations shall be developed for the drydocking facility. The manning plan shall describe each station to be manned, the responsibilities of each station, and the qualification criteria for personnel manning those stations during all drydocking operations. Modifications to the manning plan is permitted at the discretion of the Dockmaster.

10.4 VERIFICATION OF DRY DOCK CAPABILITY

The Dockmaster and/or the Dry Dock Engineer shall verify that the dry dock has the capability to drydock the vessel safely. The items to check include vessel fit, the expected work to be completed while in the dry dock, dry dock loading, and dry dock system stability (for floating dry docks and marine railways).

10.4.1 Vessel Fit Using the length, beam, and draft of the vessel, determine if the vessel will fit into and/or onto the dry dock. Minimum clearance of 1 ft (0.3 m) is required between the vessel and the dry dock wingwalls, pontoon deck, floor, sill, gate, blocks, and so forth, while bringing the vessel into or out of the dry dock. Clearance of 1 ft (0.3 m) is required between the vessel and the dry dock wingwalls, pontoon deck, floor, sill, gate, and so forth when the ship has fully landed on the blocks.

10.4.2 Expected Work The shipyard shall determine that there are sufficient space and resources available to accomplish the expected work.

10.4.3 Dry Dock Loading Using the expected vessel displacement of the vessel's drydocking condition, verify that the imposed vessel loading does not exceed the dry dock's certified rated capacity and certified rated keel load.

10.4.4 Dry Dock System Stability (Dry Dock and Vessel) For floating dry docks, the Dockmaster and/or the Dry Dock Engineer shall verify that the vessel KG is below the allowable vessel KG versus vessel displacement weight stability curve.

10.5 VESSEL DATA FORM

The shipyard shall obtain the following vessel information from the vessel owner:

1. Vessel's Drydocking Plan;
2. Vessel's longitudinal center of gravity in the drydocking condition relative to the first and last keel blocks and the vessel displacement in docking condition;
3. Vessel's hydrostatic data (curves of form, KG, etc.);
4. Vessel's drafts (forward/midships/aft);
5. Vessel's loading, soundings, and stability data;
6. Location of heavy movable items;
7. Vessel's dry dock service requirements (electrical power, steam, fresh water, salt water, cooling water, telephone, data links, sewage, air, etc.);
8. Previous drydocking reports;
9. Vessel's planned work while in dry dock;
10. Diver's report (prior to drydocking); and
11. Maximum allowable hull pressure.

10.6 DRYDOCKING CALCULATIONS

Prior to the vessel entering the dry dock, the shipyard shall perform drydocking calculations using the intended drydocking condition of the vessel using internationally recognized methods. The vessel's drafts, soundings, and loading condition shall be recorded. After the calculations are complete, no vessel changes which could affect the drydocking condition are allowed until the vessel lands on the blocks.

Drydocking calculations shall be performed by the Dockmaster and/or the Dry Dock Engineer. All personnel involved in the drydocking calculations shall possess the technical qualifications, including relevant experience, education, and training required to perform the calculations.

The calculations shown in Table 10-1 are required.

Table 10-1. Required drydocking calculations.

Required Calculations	Floating Dry Dock	Graving Dock	Marine Railway	Vertical Lift
Block Loading (10.6.1)	X	X	X	X
Dry Dock Loading (10.6.2)	X	X	X	X
Vessel Stability (10.6.3)	X	X	X	X
Vessel Trim (10.6.4)	X	X	X	X
Dry Dock Stability (10.6.5)	X			
Dry Dock Pumping Plan (10.6.6)	X			

10.6.1 Block Loading The following block loading calculations shall be computed to ensure the block loading limits are not exceeded:

1. Overall block loading,
2. Knuckle (sue) loading, and
3. Side block/shore loading.

Use Table 10-2 as the source for safe allowable wood stresses when calculating block loading.

Table 10-2. Wood properties.

Block Material	Safe Loading Limit Normal Operations (psi)	Maximum Loading Limit Overturning Moments (psi)
Douglas Fir	400	800
Red Oak	600	1300
White Oak	600	1300
Yellow Pine	300	700

10.6.1.1 Overall Block Loading Using the displacement of the vessel to be drydocked, calculate the overall block loading pressures. These values shall not exceed the block pressures of the permissible safe loading limit (psi) for the particular softwood type used on the blocks.

10.6.1.2 Knuckle Block Loading Using the displacement and trim of the vessel to be drydocked (relative to the dry dock), calculate the knuckle (sue) block loading pressures. These values shall not exceed the block pressures of the permissible safe loading limit (psi) for the particular softwood type used on the blocks.

10.6.1.3 Side Block/Shore Loading Using the expected maximum wind and seismic forces on the vessel, calculate the overturning moments and the number of side block/shores required to support the vessel. The permissible maximum loading limit is used with overturning moments.

10.6.2 Dry Dock Loading Using the shipyard's drydocking plan, calculate the bearing load on the keel and side blocks, and confirm that these loads are within the acceptable limits of the dry dock as developed in the condition assessment.

10.6.3 Vessel Stability Ensure that the vessel is stable through all phases of the docking evolution.

10.6.3.1 Draft of Instability Using the vessel's hydrostatic and loading data, calculate the draft of instability of the vessel. It is imperative that the side blocks/shores are hauled/placed into position prior to the vessel reaching the draft of instability.

10.6.4 Vessel Trim The relative trim between the vessel's keel and the plane of the keel blocks must be less than 1/100 of the vessel's length. In addition, the calculated knuckle load must be acceptable for block pressure and loading into the vessel.

10.6.5 Floating Dry Dock Stability Using the vessel's displacement and dry dock stability curve, ensure that the dry dock stability is maintained within safe limits during all phases of the drydocking evolution.

10.6.6 Floating Dry Dock Pumping Plan Using the dry dock tank capacity tables/curves, develop a pumping plan detailing the dry dock tank levels for each stage of dry dock operation. Ensure that each tank is dewatered in proportion to the load distributed above the tank. Pressing up or emptying dry dock ballast tanks to obtain adequate GM by minimizing free surface effect is not acceptable.

10.7 UNDOCKING CALCULATIONS

Prior to undocking, the vessel's undocking condition must be calculated. Starting with the as-drydocked condition, these calculations are used to predict the vessel's float-off condition, including drafts, stability, and possible list or trim. After the vessel has landed on the blocks, any weight additions, removals, or movements of 500 lb (0.227 m tons) or more affecting the vessel's center of gravity must be accounted for in these calculations. The vessel shall be ballasted or loaded so that there is acceptable trim and no calculated list. Once the undocking condition is established, no changes which could affect the undocking condition are allowed until the vessel is afloat.

For floating dry docks, dry dock stability and pumping plan calculations are required if the vessel loading has changed.

10.8 SHIPYARD'S DRYDOCKING PLAN

The shipyard shall develop a drydocking plan for the dry dock in which the vessel will be drydocked. The shipyard's drydocking plan shall contain the information in the following three sections.

10.8.1 General Notes All pertinent information shall be documented in the General Notes including

1. Explanation of drydocking positions,
2. Blocking and shoring details including size and numbering, and
3. Scale of drawing.

10.8.2 Drawing Details The Drydocking Plan shall show a plan view, profile view, and section view of the vessel on the blocks. These views shall include the following information:

1. Location of the stern reference point (SRP).
2. Location of the transverse centerline.
3. Location of baseline.
4. Distance from the SRP to the first keel and side blocks.
5. Vessel outline shall include hull openings, appendages, and protuberances that may affect the drydocking.
6. Minimum keel block height above dry dock floor. Clearances from vessel appendages and protuberances to the dry dock floor.
7. If needed, required clearances for removal of vessel components such as propellers, sonar domes, rudders, and shafts.
8. Omitted blocks shall be notated.
9. Block information.

The plan shall include the following information for all blocks:

1. Distance from SRP to after edge of the first keel block.
2. Longitudinal spacing (preferably evenly).
3. Distance from centerline of the vessel to the inboard edge of the side block.
4. Height relative to baseline height. When using fitted caps, the heights of each block shall be provided on the block's aft side at three transverse points: inboard corner, midpoint, and outboard corner. The aft to forward shape shall be given as a slope.
5. Size of bearing surface.
6. Details for stacking, cribbing, bracing, and/or other special instructions.
7. Indication and details of any dynamic blocks such as sliding, hinged, or truncated blocks, among others.

10.8.3 Hull Openings below the Waterline A table shall be generated that indicates the following information:

1. Purpose of opening(s),
2. Size of opening(s), and
3. Vertical, longitudinal, and transverse location of opening(s).

10.9 DOCUMENTATION REQUIREMENTS

10.9.1 Drydocking Report A drydocking report is required for each drydocking and undocking operation. The Drydocking Report documents the information relevant to the drydocking (Figure 10-1). The report shall, at a minimum, include the following information:

1. Name of the Dockmaster,
2. Dry dock used,
3. Name of vessel,
4. Type of vessel and class (if applicable),
5. Owner's name,
6. Shipyard's drydocking plan,
7. Reference number,
8. Drydocking position utilized,
9. Height of keel blocks (baseline) above dock floor,
10. Principal dimensions of the vessel [LOA, LBP, Beam (extreme and molded)],
11. Vessel drydocking condition
    (a) Drydocking drafts,
    (b) Drydocking displacement,
    (c) Time vessel landed on blocks, and
    (d) Height of tide at time of docking (if applicable),
12. Vessel undocking condition
    (a) Undocking drafts,
    (b) Undocking displacement,
    (c) Time vessel floated off the blocks,
    (d) Height of tide at time of undocking (if applicable),
13. Date and time vessel exited the dock,
14. Date and time vessel entered dock, and
15. Vessel work accomplished (hull painting, alterations, etc.) that will impact future drydockings.

10.9.2 Dry Dock Logbook A dry dock logbook shall be maintained at the main control station for the purpose of recording all pertinent details regarding the use and maintenance of the drydocking facility during all dry dock operation. The dry dock logbook shall include drydocking evolutions, lay-days, and periods when the dock is being maintained while not in use (Figure 10-2). The data contained within the logbook shall include the following:

1. Name of the Dockmaster;
2. Times for the commencement and completion of critical steps, such as when water enters dry dock basin/pontoon/platform;

DOCKING REPORT

Date Entered Dock ____________________ Date Exited Dock ______________

Name of Vessel ______________________________ Type & Class ________________

Owner's Name ______________________________ Flag ________________________

Builder ____

Registry – ABS ________ Lloyds ________ Veritas (Fr.) ________ Other __________

Docking Plan Available YES ☐ NO ☐ Plan Number & Location ____________________

Dock Used ______________________ Height of Keel Blocks above Floor ______________

Time Vessel Landed on Blocks ____________ Time Vessel Lifted Off Blocks ____________

Draft-in Dock (AFT) ________ (FWD) ________ Trim ________ Tide ____________

Draft-out Dock (AFT) ________ (FWD) ________ Trim ________ Tide ____________

Length Overall ____________________ Length between Perpendiculars ______________

Beam (Extreme) ______________________ Beam (Molded) ____________________

Dead Rise ______________________ Docking Position Used ______________________

Docking Displacement ________________ Undocking Displacement ____________________

NOTES ____

____

____

SIGNED ______________________________
DOCKMASTER

Figure 10-1. Example Drydocking Report.

vessel crosses the sill, vessel centered, use of divers, vessel lands on the blocks, basin/platform dry, lay period begins, start undocking operations, and so forth;
3. Tides;
4. Weather observations during the dry dock operations such as temperature, wind speed and direction, tides, and sea state;
5. Vessel and dry dock drafts (if applicable);
6. Casualties (if any);
7. Inoperable equipment;
8. Maintenance performed; and
9. Other items of interest.

10.9.3 Dockmaster Log Book All Dockmasters shall maintain a log book that will be used in part for their certification and recertification. The data contained within the log book shall include the following:

1. Date of drydocking or undocking,
2. Dry dock name and type,
3. Name of vessel, and
4. Function during evolution (Dockmaster or Dockmaster-in-training).

10.10 BLOCK BUILD REQUIREMENTS

Keel blocks and side blocks shall be built in accordance with the shipyard's drydocking plan.

10.10.1 Reference Planes Reference planes (horizontal and vertical) to which blocks are placed and built are established relative to the following:

1. SRP establishes the longitudinal reference.
2. Centerline establishes the transverse reference.
3. Baseline establishes the vertical reference.

10.10.2 Block Materials The material and general construction of the base blocks for the keel and side blocks shall be similar.

DRY DOCK LOG BOOK

DRY DOCK NAME:	
DOCKMASTER IN CHARGE:	
	TIMES OF EVENTS:
TIDE:	
WEATHER CONDITIONS:	
DOCK DRAFTS:	
CASUALTIES:	
INOPERABLE EQUIPMENT:	
MAINTENANCE PERFORMED:	
OTHER:	

Figure 10-2. Example dry dock log book.

Most of the vessel's load shall be on the keel blocks. In no case shall the side blocks be of stiffer construction than the keel blocks.

10.10.3 Block and Cradle Condition Block's and cradle's ability to support designed load shall not be impaired. Items that may impact a block's/cradle's ability to carry load are cracks, spalling, broken welds, structural damage, corrosion, or other signs of significant damage. Items which may impact wooden blocks are hollows, cracks, crushing, rot, or other signs of significant damage. The wood fastening system (fasteners, battens, dogs, banding, etc.) shall be checked to ensure that the wood will not float or move when submerged.

10.10.4 Soft Caps Softwood caps shall form the top layer of keel and side blocks. The function of softwood is to crush slightly under load to comply with any irregularities in the blocks or vessel hull. Soft caps shall be no less than 2 in. (50 mm) in thickness. In no case shall the keel soft caps be thicker than those on the side blocks. In cases of narrow keels or high expected loads, the softwood caps shall be placed under hardwood or metal plates used to distribute the load. The wood grain of soft caps shall be oriented perpendicular to the vessel's strength member the block supports (keel, longitudinal, frame, etc.) Softwood caps that show signs of excessive crushing, cracks, warping, rot, or other material defects shall be replaced. The soft cap fastening system (fasteners, battens, dogs, banding, etc.) shall be checked to ensure that soft caps will not float when submerged.

10.10.5 Block Position Tolerance The keel and side block caps shall be within 1/4 in. (6 mm) of height dimensions, 1/2 in. (13 mm) of the transverse distances, and 1 in. (25 mm) of the longitudinal distances specified on the shipyard's drydocking plan.

10.10.6 Line of Force The line of action of the force normal to the vessel's hull for all blocking must pass through the middle third of each layer of the block build.

10.10.7 Block Stability To ensure block stability, the ratio of height to width (or length) must not exceed 2.0 for all uncribbed individual blocks. The ratio of uncribbed height to width (or length) is called aspect ratio. Blocks exceeding this ratio in the transverse direction must be stacked pyramid style, cribbed, or braced.

DOCKMASTER LOGBOOK

DATE	DOCKING UNDOCK OTHER	DOCK USED	VESSEL NAME	FWD DRAFT	AFT DRAFT	ASSISTING OR IN CHARGE

Figure 10-3. Example Dockmaster log book.

When required, pyramid blocks will be constructed so that the block bases are fully supported so that no blocks "overhang" the blocking on which it rests. The pyramid method shall increase the block base to achieve an aspect ratio of 2.0 or less. When required, cribbing shall be installed at the forward and after 1/3 of the ends of the line of blocks and shall include four blocks on either end at a minimum. Cribbing shall consist of 3 in. (75 mm) minimum thick hardwood, or steel beams that connect a layer of blocking material in the longitudinal direction. Cribbing shall be installed as either two-column cribbing or continuous cribbing; two-column cribbing must pass from the aft most side of one block to the forward most side of the next block. Continuous cribbing will join multiple blocks together with the cribbing joints being centered over blocks. In all cases, cribbing voids of more than 3 in. (75 mm) between timbers will not be tolerated.

When required, bracing of blocks shall be made to provide adequate overturning resistance in accordance with the drydocking calculations.

10.10.8 Verification After completion of the block build, the Dockmaster and another individual familiar with block construction shall each independently check the following in accordance with the shipyard's drydocking plan:

1. Location of blocks,
2. Heights of blocks,
3. Blocks will not interfere with hull openings and projections,
4. Alignment of the transverse centering system and longitudinal positioning system, and
5. Any other issues which can affect the safe landing of the vessel on the blocks.

CHAPTER 11
DRY DOCK OPERATIONS—REQUIREMENTS COMMON TO ALL DRY DOCK TYPES

This section defines the operational and documentation requirements common to all dry dock types to safely drydock and undock vessels. Prerequisite procedures and checks are required prior to commencing drydocking/undocking operations. All dry dock operating procedures shall be prepared sequentially, in step-by-step detail. The documented procedures are a matter of record and shall be strictly followed. The sequential checklists provided in Appendix C may be used to develop facility specific procedures that shall be followed during drydocking/undocking operations.

11.1 DRYDOCKING/UNDOCKING CONFERENCE

Prior to vessel drydocking or undocking, a conference for the exchange of information shall be held to clarify operational plans. Attendees shall include, at a minimum, key shipyard personnel, the Dockmaster, key vessel personnel including the Master, Chief Mate, and Engineer, and the tugboat Captain and/or Pilot (if involved). The conference may be convened by teleconference if it is not practical to conduct the conference in person. The meeting shall cover at least the following topics:

1. Key personnel involved;
2. Pertinent information and schedule of activities;
3. Details on facility access including security, parking, and so forth;
4. Weather forecast and limits;
5. Communications plan;
6. Operational plan (vessel movement, tugs, fenders, line handlers, capstans, hauling blocks, etc.);
7. Vessel expected condition at arrival (drafts, list, trim, ballast requirements, propeller, rudder, status of underwater retractable equipment housed, etc.);
8. Shore services and gangways;
9. Use of divers;
10. Vessel work planned;
11. Other items that involve safety (man overboard, personnel protective equipment, fall protection, etc.).

11.2 PRIOR TO DRYDOCKING OR UNDOCKING OPERATIONS

11.2.1 Operating Procedures Dry dock operating procedures shall be available at all stations during the drydocking operations.

11.2.2 Communication Verify all methods of communication are working properly.

11.2.3 Safety Meeting and Assignment of Personnel A safety meeting shall be convened prior to dry dock operations to instruct drydocking operations personnel with specific safety concerns. Personnel assignments and details of the drydocking operation will be discussed at this meeting.

11.2.4 Dry Dock Systems and Equipment Verify that all dry dock systems and equipment are ready for dry dock operations, including lines, fenders, wire ropes, in haul systems, capstans, pumps, valves, control systems, power systems, air systems, and water systems.

11.2.5 Final Dry Dock Checks Prior to flooding/lowering the dry dock, the Dockmaster shall check all dry dock blocks to ensure that all blocks and wood are properly secured, cranes and other topside items are secured/clear for vessel entry, and the dry dock is ready for flooding/lowering.

11.2.6 Manning Before starting dry dock operations, the Dockmaster shall ensure that all stations are properly manned. The Dockmaster shall be in a location where he or she can see the vessel's movements and control them. The Dockmaster shall be stationed on the dry dock or shore (not on a vessel while it is in motion into the dry dock).

11.2.7 Weather All dry docks shall have applicable weather limitations listed in their operating procedures. In no case shall operations be conducted in the event of blinding precipitation, dense fog, localized lighting strikes or winds, tides, or currents that exceed the limitations of the dry dock. Weather conditions shall be assessed by the Dockmaster prior to and during drydocking operations. The Dockmaster shall impose stricter weather limitations as needed. For example, vessels with large sail areas will have a lower maximum acceptable wind limit than vessels with small sail areas.

11.3 DRYDOCKING OPERATIONS

11.3.1 Vessel Condition Prior to the vessel entering the dry dock, the Dockmaster shall verify the vessel's draft readings, list, trim, retractable equipment, propeller(s), rudder(s), sea suctions, cathodic protection, and so forth. This is to ensure that the vessel's condition afloat is safe for drydocking in accordance with the shipyard drydocking plan and drydocking calculations.

11.3.2 Minimum Clearance While bringing the vessel into the dry dock, 1 ft (0.3 m) clearance is required between the vessel hull or projections and the dry dock structure (wingwalls, pontoon deck, floor, sill, gate, blocks, etc.). The vessel should not come in contact with the dry dock structure while bringing the vessel into the dry dock. At any anticipated points of contact, fenders or rollers shall be used to keep the vessel one foot off the dry dock structure.

Clearance of 1 ft (0.3 m) is required between the vessel and the dry dock wingwalls, pontoon deck, floor, sill, gate, and so forth, when the ship has landed on the blocks. Once the vessel has landed on the blocks, there shall be a minimum of 1 ft (0.3 m) clearance between the hull opening or projections and the dry dock structure or blocks.

11.3.3 Transition from Ship or Tug Propulsion to Handling by Dry Dock Lines When the vessel is a safe distance from the dry dock, heaving lines shall be passed between the dry dock and the vessel to make up the handling lines/wire ropes fast to the vessel. It may be necessary to stop the vessel's movement to allow tugs to hook up/cast off their lines.

Tugs may continue to be used to assist with the movement of the vessel into dry dock after their lines have been cast off. They may position themselves at the entrance to the dry dock such that they prevent loss of control of the vessel until an adequate number of dry dock handling lines/wire ropes have been connected to the vessel.

Warning: The use of vessel or tug propulsion within the dock should be used with caution because propulsion wash can disturb or damage the blocks and/or the dry dock itself (e.g., granite block mortar joints).

11.3.4 Exchange of Responsibility and Operational Control As the leading edge of the vessel to be drydocked crosses the dry dock sill (or opening) and the ship is fair to be drydocked, the Dockmaster shall assume operational control of the vessel and take full responsibility for the safety of the vessel being drydocked. The sill time (or crossing of the opening) is to be made a matter of record and recorded in the ship and dry dock logs.

11.3.5 Positioning the Ship The ship shall be hauled into the dry dock, and centered transversely and longitudinally over the blocks according to the shipyard drydocking plan. This centered position shall be maintained until the vessel is landed on the blocks.

11.3.6 Alignment over the Blocks Alignment devices, such as centering bobs, chains, battens, transits, or other means of ensuring position, shall be used to align the ship within the dry dock.

11.3.7 Landing the Ship

11.3.7.1 Vessel Drafts and Dry Dock Water Levels When the vessel has been positioned longitudinally and transversely in the dry dock and is fair for landing, the final prelanding drafts of the vessel and water levels in the dry dock shall be recorded. The Dockmaster shall monitor both continuously while landing the vessel to ensure it lands when anticipated. If the vessel lands too early or does not land as anticipated, operations shall stop until the source of the error has been determined. It may be necessary to refloat the vessel and/or send divers down to investigate the matter. The times of when the vessel first touches the blocks, when the vessel is fully landed on all blocks, and when the final haul block or shore is installed shall be recorded.

11.3.7.2 Line Handling All line handlers shall remain on station and be prepared to handle all lines until the vessel is landed on all blocks. As the vessel nears landing, the lines to the vessel may become slack or taut, and it will require constant working of the lines to hold the vessel in position.

11.3.7.3 Hauling Side Blocks When hauling side blocks are used, they shall be hauled prior to the vessel reaching its draft of instability (GM = 1 ft or 0.3 m). The draft at which the side blocks will be hauled shall be determined as part of the drydocking calculations.

After the vessel initially lands fair on the keel blocks and it is determined that its transverse and longitudinal positions are acceptable, the vessel shall be lowered/lifted to a calculated stopping draft. At that point, the hauling side blocks shall be hauled into position and secured.

11.3.7.4 Shoring If shores are used instead of side blocks, placement of the shores shall be accomplished prior to the vessel reaching its calculated draft of instability (GM = 1 ft or 0.3 m). Shoring installed in conjunction with side blocks can be placed any time after the vessel has landed fair on the keel blocks.

11.3.8 Stopping Drydocking Operations The drydocking of the vessel shall be stopped by the Dockmaster and the vessel refloated if

1. Vessel is misaligned longitudinally greater than 3 in. (75 mm).
2. Vessel is misaligned on the keel blocks greater than 3 in. (75 mm) transversely.
3. Vessel takes an unexpected list at landing.
4. Vessel indicates instability or near instability.
5. Vessel lands at a dry dock draft greater or less than predicted, which may indicate a foreign object on the block line, an unknown protrusion below the vessel, or damage to the blocks.
6. Less than 70% of the expected block contact is made.
7. Shore buckles (if used).
8. Any condition in which the safety of the ship, the dry dock, or personnel are in danger.

11.3.9 Divers Divers can assist with drydocking operations. If divers are used during a drydocking evolution, divers shall be briefed prior to entering the water. The briefing shall include a review of the Drydocking Plan, noting locations of suction intakes, as well as a walk-through of the dry dock and the block build. Divers can be available to assist with the following operations:

1. To verify that the vessel is in the correct position relative to the block build and has the required clearance.
2. To resolve underwater discrepancies or obstructions.
3. To check the vessel's proper alignment and fit on the blocks.
4. To assist in the rigging of hauling blocks, hinged blocks, shores or other special drydocking operational situations.

11.3.10 Grounding Cables, Gangways, and Services Grounding cables or straps designed to protect the vessel from the effects of welding and electrical storms shall be attached at the direction of the Dockmaster after the vessel has landed safely on the blocks. Gangways and service lines shall be installed after the vessel has landed on the keel and side blocks/shores are secured in position.

11.3.11 Final Liquid Load Readings As directed by the Dockmaster, after the vessel has landed and before any vessel weight changes (weight additions, removals, or shifts) occur, the vessel shall take and record its final soundings and record locations of heavy equipment and objects.

11.3.12 Examination of the Blocks/Cradles/Shores After removal of water off the dry dock floor, a visual inspection of the vessel shall be made to confirm correct positioning on the blocks, cradles, and shores. All blocks, cradles, and shores shall be inspected for acceptable loading and stability. All corrections to the blocks, cradles, and shores shall be made as determined by the Dockmaster.

If there is excessive strain on the vessel or the vessel has been landed outside of acceptable limits, it will be necessary to add additional support or refloat the vessel.

11.4 LAY PERIOD

11.4.1 Corrections to the Vessel Blocking Plan During the lay period, any discrepancies with the Vessel Blocking Plan and/or modifications to the vessel that affect the Vessel Blocking Plan shall be noted on a redline drawing. Copies of these updated plans will be retained by the shipyard and provided to the vessel owner. Any modifications to the vessel that affect the subsequent undocking of the vessel shall be recorded and submitted to the Dockmaster prior to undocking.

11.4.2 Weight Changes The shipyard shall maintain a record of all vessel weight changes (additions, removals, moves) over 500 short tons (454 metric tons) during the lay period. This information shall be submitted to the Dockmaster prior to undocking for use in performing the undocking calculations. Permanent weight changes that affect the light ship condition and stability of the vessel shall be submitted to the vessel owner.

11.4.3 Watertight Integrity Temporary closure plates shall be available prior to removing plates or cutting access openings below the waterline. These closure plates shall be used for emergency sealing of temporary access openings. Removal of the temporary closures can be accomplished after the potential breach of watertight integrity no longer exists.

11.5 UNDOCKING OPERATIONS

11.5.1 Vessel Condition Prior to floating the vessel, the vessel shall be loaded to a safe condition based on the undocking calculations. The vessel's calculated undocking stability, displacement, drafts, trim, and list shall be acceptable. If possible, matching the vessel's drydocking loading condition is desired. Verify that retractable equipment, propeller(s), rudder(s), sea suctions, cathodic protection, and so forth are in an acceptable condition. This is to ensure that the vessel's condition is safe for undocking in accordance with the shipyard drydocking plan and undocking calculations.

11.5.2 Operating Procedures The undocking procedures shall be available at all stations during the undocking operations.

11.5.3 Communication Verify that all methods of communication are working properly.

11.5.4 Safety Meeting and Assignment of Personnel A safety meeting shall be convened prior to dry dock operations to instruct drydocking personnel with specific safety concerns. Personnel assignments and details of the drydocking operation will be discussed at this meeting.

11.5.5 Dry Dock Systems and Equipment Verify that dry dock systems and equipment are ready for dry dock operations, including lines, fenders, wire ropes, in haul systems, capstans, pumps, valves, control systems, power systems, air systems, water systems, and so forth.

11.5.6 Final Dry Dock Checks Prior to flooding/lowering the dry dock, the Dockmaster will check all dry dock blocks to ensure that all blocks and wood are properly secured, the dry dock meets environmental requirements, and is ready for flooding/lowering. While a vessel is in motion in a dry dock, the Dockmaster shall be in a location where he or she can see the vessel's movements. The Dockmaster shall not be on a vessel while leaving the dry dock.

11.5.7 Manning Before starting dry dock operations, the Dockmaster shall ensure that all stations are properly manned.

11.5.8 Weather All dry docks shall have applicable weather limitations listed in their operating procedures. In no case should operations be conducted in the event of blinding precipitation, dense fog, localized lightning strikes or winds, tides, or currents that exceed the limitations of the dry dock. Weather conditions shall be assessed by the Dockmaster prior to and during drydocking operations. The Dockmaster shall impose stricter weather limitations as needed. For example, vessels with large sail areas will have a lower maximum acceptable wind limit than vessels with small sail areas.

11.5.9 Vessel Condition Prior to refloating the vessel, verify the vessel's retractable equipment, propeller(s), rudder(s), sea suctions, cathodic protection, and so forth are in an acceptable condition. This is to ensure that the vessel's condition afloat is safe for undocking in accordance with the specified drydocking plan and undocking calculations.

11.5.10 Minimum Clearance While bringing the vessel out of the dry dock, 1 ft or 0.3 m) clearance is required between the vessel hull or projections and the dry dock structure (wingwalls, pontoon deck, floor, sill, gate, caisson, blocks, etc.). The vessel should not come in contact with the dry dock structure while bringing the vessel out of the dry dock. At any anticipated points of contact, fenders or rollers shall be used to keep the vessel 1 ft off the dry dock structure.

11.5.11 Divers Divers can assist with undocking operations. If divers are used during an undocking evolution, divers shall be briefed prior to entering the water. The briefing shall include a review of the shipyard drydocking plan, noting locations of suction intakes. Divers can be available to assist with the following operations: (1) To resolve underwater discrepancies or obstructions and (2) to assist in the derigging of hauling blocks, hinged blocks, shores, or other special drydocking operational situations.

11.5.12 Flooding Checks Stop the water level at 2 ft (.610 m) before the vessel expected floating draft. Conduct a thorough inspection of the underwater hull of the vessel looking for any leaks or flooding. Any leaks or flooding problems shall be corrected or the Vessel Master and the Dockmaster shall agree that the condition is not serious before proceeding with the undocking.

11.5.13 Grounding Cables, Gangways, and Service Lines Grounding cables or straps, gangways, and service lines shall be removed prior to floating the vessel.

11.5.14 Floating the Vessel

11.5.14.1 Line Handling All line handlers shall be on station and be prepared to handle all lines prior to floating the vessel. As the vessel floats, the vessel lines shall be rigged to hold the vessel in position. The lines to the vessel may become slack or taut, and it will require constant working of the lines to hold the vessel in position.

11.5.14.2 Vessel Drafts and Dry Dock Water Levels While floating the vessel, the water levels in the dry dock and the vessel's drafts shall be monitored by the Dockmaster to ensure that the vessel floats when anticipated. If the vessel floats too early or not when anticipated, operations shall stop until the source of the error has been determined. The times of when the vessel first floats free of any blocks and when it is fully floating free of all the blocks shall be recorded. The final vessels floating drafts and the final water level in the dry dock shall be recorded prior to the vessel exiting the dry dock.

11.5.14.3 Hauling Blocks and Shores Hauling blocks shall be hauled out of the way after the vessel has floated off the blocks. Shores shall be removed at the same draft as when they were put into position and no later than draft at which the ship floats off the blocks.

11.5.15 Transition from Handling Lines to Ship or Tug Propulsion As the ship is hauled out of the dry dock, tugs may be used to assist with the movement and control of the vessel. It may be necessary to stop the vessel's movement to allow the tugs to be made-up to the vessel. When the vessel is a safe distance from the dry dock, lines/wire ropes shall be cast off or taken in.

11.5.16 Exchange of Responsibility and Operational Control As the aftermost edge of the vessel crosses the dry dock sill (or opening) and the ship is fair, the Dockmaster relinquishes operational control of the vessel to the tugboat Captain, Pilot, or Master of the vessel. The sill time is to be made a matter of record and recorded in the ship and dry dock logs.

11.5.17 Stopping Undocking Operations The undocking of the vessel shall be stopped by the Dockmaster if

1. Vessel takes an unexpected list when floated.
2. Vessel floats at a dry dock draft greater or less than predicted.
3. Vessel indicates instability or near instability.
4. Any condition in which the safety of the ship, the dry dock, or personnel are in danger.

11.5.18 Accounting for the Dry Dock's Blocks/Cradles/Shores The Dockmaster shall account for all of the blocks/cradles/shores once the undocking evolution is complete. It is possible that during paint operations or extremely low temperatures, wood may adhere to the vessel during the undocking. If wood is found to be missing, the vessel's personnel shall be notified that divers will be required to check the underwater hull to remove the adhered wood.

Any moved blocks/cradles/shores can be an indication that the vessel hit them during the undocking. If blocks/cradles/shores are found to be moved, the vessel's personnel shall be notified, and divers will be required to inspect the underwater hull.

CHAPTER 12
DRY DOCK OPERATIONS—REQUIREMENTS UNIQUE BY DRY DOCK TYPE

This chapter defines the operational requirements, unique by dry dock type, to safely drydock and undock vessels.

12.1 FLOATING DRY DOCK SPECIFIC OPERATIONS

This section describes floating dry dock specific requirements not included in Chapters 10 and 11 of this standard.

12.1.1 Operating Procedures All floating dry dock operating procedures shall be prepared sequentially, in step-by-step detail. The documented procedures are a matter of record and shall be strictly followed. Prerequisite procedures shall be completed prior to commencing the drydocking evolution, including but not limited to, cycling valves, operating pumps, and testing tank water level systems.

12.1.2 Baseline Height The baseline height plane shall be established with the dry dock in an unstressed (unloaded) condition.

12.1.3 Pumping Plan The purpose of the pumping plan is to provide guidance on water levels in the tanks while ballasting and deballasting the dry dock. The pumping plan shall include water levels in each tank at any intermittent stopping points. Intermittent stopping points include, but are not limited to, dry dock submergence, prior to landing diver checks, hauling side blocks (if applicable), block fit checks by divers, dry dock final pumped position, and watertight checks (during undocking).

12.1.4 Dry Dock Deflection Dry docks, which utilize a deflection monitoring system, shall provide information on the longitudinal bending of the dry dock. The acceptable bending limits of the dry dock shall not be exceeded.

12.1.5 Dry Dock Stability Verify and document vessel KG is below the allowable vessel KG versus vessel displacement weight stability curve for all phases of lift. Figure 12-1 shows a sample KG versus weight curve. This curve is used by the Dockmaster to verify adequate stability of the vessel/dry dock combination when drydocking/undocking any vessel. If the plot of the vessel's drydocking/undocking displacement and its KG intersect on or below the line, the dry dock has sufficient stability to lift the vessel and maintain the minimum GM for the dry dock and vessel system in accordance with this standard. If the plot of the vessel's drydocking/undocking displacement and its KG intersect above the line, the dry dock does not have sufficient stability to lift the vessel and the GM for the dry dock and vessel system will be below the minimum required and may possibly be negative, indicating an unstable condition.

12.1.6 Dry Dock Operational Depths Floating dry docks shall maintain 1 ft (0.3 m) clearance from the bottom of the bay or river, unless designed to land on the bottom.

12.1.7 Dry Dock List and Trim When the dry dock is at operating draft, pier side list and trim shall be maintained near zero. Sometimes the dry dock will be kept at a slight list/trim to facilitate water runoff.

While the pontoon deck is transiting through the water plane, trim the dry dock slightly to stabilize the dry dock. The trim effectively controls the speed at which the pontoon deck becomes flooded/awash, and the ship-dock system loses stability.

The dry dock trim shall match the vessel's trim when landing/undocking the vessel on the dry dock. The maximum trim of the floating dry dock shall be less than 1/100 of the dry dock's length.

12.1.8 Dry Dock Drafts Record dry dock drafts at critical points in the dry dock operations. These include, but are not limited to, dry dock submergence, prior to landing diver checks, hauling side blocks (if applicable), block fit checks by divers, dry dock final pumped position, and watertight checks (during undocking).

12.1.9 Cranes All dry dock cranes shall be secured in their resting positions during dry dock ballasting and deballasting operations. Cranes may be used to remove gangways, service lines, and grounding cables when the dry dock is not ballasting and deballasting.

12.1.10 Severe Weather Protect the dry dock by lowering, moving, landing, and/or securing in position. Protect the dry

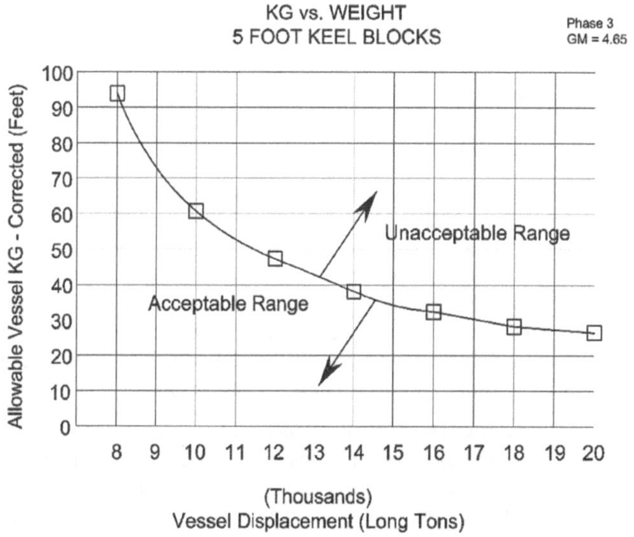

Figure 12-1. Sample vessel KG versus vessel displacement weight curve.

dock by moving near-by vessels and floating objects. Prepare for the possibility that vessels in the dry dock may float off the blocks.

12.2 GRAVING DRY DOCKS SPECIFIC OPERATIONS

This section describes graving dry dock specific requirements not included in Chapters 10 and 11 of this standard.

12.2.1 Operating Procedures All graving dry dock operating procedures shall be prepared sequentially, in step-by-step detail. The documented procedures are a matter of record and shall be strictly followed. Prerequisite procedures shall be completed prior to commencing the drydocking evolution including, but not limited to, cycling valves and operating pumps.

12.2.2 Flooding the Graving Dock Flooding the dry dock is typically accomplished by allowing water to travel through a sluice gate from a flooding chamber or through flood valves in the closure gate. With free communication between the outside water and the dry dock, water will continue to flow into the dry dock until the water levels equalize. Care shall be taken during the initial flooding of the basin to minimize turbulence that could disrupt blocking, floor trench covers, and gratings. When flooding through closure gate flood valves, flood the dry dock at a reduced flood rate until a sufficient cushion of water has accumulated in the basin to minimize turbulence.

12.2.3 Specific Gate Operation Requirements There are numerous types of entrance closures utilized for graving dry docks such as hinged gates, rolling gates, caisson gates, and miter gates. All types of closures shall have specific procedures available that include operating instructions and limitations.

12.2.3.1 Floating Caisson Gates When the inside and outside water levels of the dry dock have equalized, the caisson gate shall be deballasted to allow the gate to float out of the dry dock's sill. Attempting to deballast the gate prior to water level equalization will cause an undesirable lifting of the gate that must be avoided.

After the ship has been safely hauled into the dry dock and is secured in the dry dock, the gate shall be placed back in position and secured before any further operations begin. The Dockmaster shall ensure that the gate is installed properly. The caisson gate shall be fully ballasted into position.

The Dockmaster shall witness the caisson gate's condition during ballasting/deballasting and correct trim greater than 6 in. (152 mm) to avoid damage to the dry dock sill.

Tugs and/or line handlers ashore control the movement of the gate to a location that does not interfere with the subsequent vessel movement into/out of the dry dock.

12.2.3.2 Hinged Gates When the inside and outside water levels have equalized, the hinged gate shall be ballasted (if it is fitted with a ballast tank) so that it can submerge until it rests on the bottom or submerged supports. Positive verification to ensure that it is fully down and to ensure clearance for the subsequent operation of the dry dock is required. With the ship secured, the gate shall be deballasted (if fitted with a buoyancy tank) and then hauled by capstan or winch back into position.

12.2.4 Securing Dry Dock Flood Valves After the gate is secured in position and the vessel has been positioned, all flood valves shall be secured. Consideration shall be given to the possibility that the gate may lose its seal if the tide is falling. In general, the gate is not fully sealed until the inside water level is lower than the outside water level. The hydrostatic pressure will force the gate to seat into its final position.

12.2.5 Dewatering the Dry Dock Many graving dry docks will require the monitoring of ground water levels behind the dry dock walls before, during, and after dry dock pump down to prevent overstressing the dry dock walls. When operating this type of dry dock, all procedural limitations regarding groundwater levels versus water in dry dock levels shall be followed. Failure to do so could lead to dry dock wall collapse.

12.2.6 Vessel List and Trim The vessel to be drydocked or floated off shall be ballasted to the near zero list and trim to minimize knuckle loads. The trim of the vessel shall not exceed 1/100 of the length of the vessel.

12.2.7 Severe Weather Protect the caisson by moving nearby vessels and floating objects. Prepare for flooding in the graving dock. Prepare for vessels in the graving dock to float off the blocks.

12.3 MARINE RAILWAY SPECIFIC OPERATIONS

This section describes marine railway specific requirements not included in Chapters 10 and 11 of this standard.

12.3.1 Operating Procedures All marine railway operating procedures shall be prepared sequentially, in step-by-step detail. The documented procedures are a matter of record and shall be strictly followed. Prerequisite procedures are required prior to commencing the drydocking evolution including, but not limited to, divers checking for debris on track, clearing of grit from track and rollers/wheels, and checking for proper tension in hauling and backing chains prior to carriage/cradle movement.

12.3.2 Drydocking Procedure

12.3.2.1 Lowering the Carriage/Cradle to Full Depth If the drydocking cradle has been in the up position for some time, or severe weather has occurred while the drydocking cradle is in the up position, divers shall swim the tracks to ensure that they are clear of debris and that an adequate length of rollers are available for the cradle. Silt and debris can accumulate on the tracks and derail the drydocking cradle.

If the drydocking cradle is not moving smoothly down the tracks or it appears that the winch is pulling the drydocking cradle down the tracks, then there is a possibility that debris on the tracks may cause the drydocking cradle to derail. The evolution shall be stopped, and divers used to survey the tracks to determine the cause of the obstruction.

The vessel shall not be brought into the marine railway until full depth position of the drydocking cradle is achieved. Full depth shall be considered a depth at which there is at least 1 ft (0.3 m) of clearance between the hull of the vessel or projections and the nearest block.

12.3.2.2 Hauling the Drydocking Cradle When the vessel is fair for landing and all line handlers are on station, the Dockmaster shall instruct the winch operator to commence hauling the cradle.

If the drydocking cradle is not moving smoothly up the tracks, then there is a possibility that debris on the tracks may cause the drydocking cradle to derail. The evolution shall be stopped, and divers used to survey the tracks to determine the cause of the obstruction.

12.3.3 Undocking Procedure When the vessel is safe for float-off and all line handlers are on station, the Dockmaster shall instruct the railway operator to lower the drydocking cradle. The drydocking cradle shall be hauled out to the full submergence position prior to the vessel being moved out of the carriage/cradle. Full submergence shall be considered a depth at which there is at least 1 ft (0.3 m) of clearance between the hull of the vessel or projections and the nearest block.

If the drydocking cradle has been in the up position for some time, or severe weather has occurred while the drydocking cradle is in the up position, divers shall swim the tracks to ensure that they are clear of debris. Silt and debris can accumulate on the tracks and derail the carriage/cradle.

If the drydocking cradle is not moving smoothly down the tracks or it appears that the winch is pulling the drydocking cradle down the tracks, then there is a possibility that debris on the tracks may cause the drydocking cradle to derail. The evolution shall be stopped, and divers used to survey the tracks to determine the cause of the obstruction.

12.3.4 Vessel List and Trim The vessel to be drydocked or floated off shall be ballasted to the near zero list and trim to minimize knuckle loads. The trim of the vessel shall not exceed 1/100 the length of the vessel.

12.3.5 Severe Weather Haul and secure the drydocking cradle in the fully hauled position.

12.4 VERTICAL LIFT SPECIFIC OPERATIONS

This section describes vertical lift specific requirements not included in Chapters 10 and 11 of this standard. The following requirements assume that the vertical lift system has a transfer system integrated into the facility. They assume that transfer cradles will be linked together, and that keel and side blocks are installed on top of the transfer cradles. Note: Vessel transfer systems are not covered by this standard.

12.4.1 Operating Procedures All vertical lift operating procedures shall be prepared sequentially, in step-by-step detail. The documented procedures are a matter of record and shall be strictly followed. Prerequisite procedures shall be completed prior to commencing the drydocking evolution including, but not limited to, control system inputs, positioning and securing the transfer cradle on the vertical lift platform, inspection of hoist system for debris or damage, and inspection of the transfer system track or roadway.

12.4.2 Drydocking Procedure When the vessel is fair for landing and all line handlers are on station, the Dockmaster shall instruct the Control Operator to commence lifting the platform. Most vertical lifts have an articulated platform with load cells on each hoist. The load monitoring system will indicate on the Control Operator's display when the vessel has sued on the knuckle block. If any increasing hoist load is indicated elsewhere on the platform or before the predicted docking draft is reached, the Dockmaster shall instruct the Control Operator to stop the lifting operation until the source of the loading error has been determined.

12.4.3 Undocking Procedure The vessel shall be undocked in accordance with the shipyard's undocking plan. The undocking procedures shall be prepared in a formal manner as a matter of record and made available at all stations during the undocking evolution.

12.4.3.1 Lowering the Platform When the vessel is safe for undocking (launching) and all line handlers are on station, the Dockmaster shall instruct the Control Operator to lower the platform. If the area below the vertical lift has been obstructed or silted in, the load monitoring system will indicate that the affected hoist loading is reducing prematurely. The lowering operation shall be stopped, and the cause of the reducing hoist load investigated.

12.4.3.2 Lowering the Platform to Full Submergence Full submergence shall be considered the depth at which there will be at least 1 ft (0.3 m) of clearance between the hull of the vessel or hull projections and the nearest block. This minimum clearance shall be maintained before and after the side blocks have been pulled away from the hull and when the vessel is being moved out of the vertical lift dry dock.

12.4.4 Vessel List and Trim The vessel to be drydocked or floated off shall be ballasted to the near zero list and trim to minimize knuckle loads. The trim of the vessel shall not exceed 1/100 of the length of the vessel.

12.4.5 Severe Weather If the vertical lift platform is empty, fully lower the platform to minimize the movement of the platform owing to wind and waves. If a vessel is on the platform, it is recommended that the vessel be undocked, secured away from the vertical lift piers, and the platform fully lowered.

CHAPTER 13
SHIPYARD RECORD-KEEPING REQUIREMENTS

13.1 PURPOSE

The shipyard shall keep and maintain files as mandated and outlined by this standard.

13.2 GENERAL RECORDS

The following shall be kept on file and maintained by the shipyard for the life of the facility:

1. Dry dock as-built drawings and specifications;
2. Dry dock capacity calculations;
3. Dry dock stability calculations (for floating dry docks and marine railways);
4. Dry dock operating limitations;
5. Documentation of dry dock repairs and/or modifications;
6. Most recent hydrographic survey;
7. Line, grade, and gauge tolerances and the most recent survey (for marine railways);
8. Dry Dock Capacity Certificate; and
9. For dry docks that do not have an installed fire protection system compliant with Occupational Safety and Health Administration (OSHA) regulations, secure and maintain a *Memorandum of Agreement* with a local fire department ensuring that the fire department can arrive at the facility within 15 min of receiving the alarm. The *Memorandum of Agreement* shall be dated within 5 years of the date of the Dry Dock Capacity Certificate.

13.3 CONDITION ASSESSMENT REPORT

The following shall be kept on file and maintained by the shipyard for the life of the facility:

1. Dry Dock Condition Assessment Report,
2. Qualifications and contact information of personnel conducting condition assessment, and
3. Engineering justification for deferred inspection items.

13.4 CERTIFICATION INSPECTION RECORDS

1. Latest copy of the certification inspection report, and
2. Qualifications and contact information of personnel conducting certification inspections.

13.5 DRY DOCK OPERATIONS

The following shall be kept on file and maintained by the shipyard:

1. Qualifications of dry dock operations personnel,
2. Qualifications and certification of the Dockmaster,
3. Dry dock manning plan,
4. Dry dock operating procedures,
5. Docking/undocking calculations,
6. Drydocking reports,
7. Dry dock log book, and
8. Dockmaster log book.

APPENDIX A
CERTIFICATION INSPECTION CHECKLISTS

APPENDIX A1 CERTIFICATION INSPECTION CHECKLIST—FLOATING DRY DOCKS

Facility Name: ___________

Dry Dock Designation: ___________

Items Inspected	Condition					Remarks
	U	M	NA	NI	S	
Mechanical and Electrical						
Dewatering/Flooding Systems						
Main Dewatering Pumps						
Motors for Dewatering Pumps						
Motor Controllers						
Lubrication						
Piping						
Drainage Pumps						
Suction Valves and Valve Operators						
Discharge Valves and Valve Operators						
Flood Valves and Valve Operators						
Sluice Valves and Valve Operators						
Crossover Valves and Valve Operators						
Tank Level Indication System						
Draft Indication System						
Deflection Detection System						
Inclinometers						
Power Systems						
Engine Generator Sets						
Shore Power for Main Power Source						
Electrical Power Distribution System						
Shore Power for Backup Power Source						
Communication Systems						
Sound Powered Phones and Radios						
Dial Telephone						

continues

(Continued)

Items Inspected	Condition U	M	NA	NI	S	Remarks
Public Address System						
Fire Alarms						
Fire Protection Systems						
Fire Pump						
Fire Main						
Fire Stations—Hoses, Nozzles, etc.						
CO_2 Extinguishers						
Dry Chemical Extinguishers						
Miscellaneous						
Alarms—Flooding, etc.						
Lighting for Operations and Security						
Deck and Breasting Winches						
Capstans/Winches/Trolleys						
Block Handling Systems						
Crane Stops and Securing Systems						
Stern and Bow Closure Machinery						
Cathodic Protection Systems						
Air Compressors and Distribution Systems						
Steel Hull Structure						
Pontoon						
Pontoon Deck – Plating						
Pontoon Deck – Stiffeners						
Pontoon Deck – Girders						
Pontoon Bottom – Plating						
Pontoon Bottom – Stiffeners						
Pontoon Bottom – Girders						
Pontoon Sides – Plating						
Pontoon Sides – Longitudinals						
Pontoon Sides – Verticals						
Long. CL Bulkheads – Plating						
Long. CL Bulkheads – Longitudinals						
Long. CL Bulkheads – Verticals						
Long. Off CL Bulkheads – Plating						
Long. Off CL Bulkheads – Longitudinals						
Long. Off CL Bulkheads – Verticals						

continues

(Continued)

Items Inspected	Condition					Remarks
	U	M	NA	NI	S	
Trans. End Bulkheads – Plating						
Trans. End Bulkheads – Horizontals						
Trans. End Bulkheads – Verticals						
Trans. Bulkheads – Plating						
Trans. Bulkheads – Horizontals						
Trans. Bulkheads – Verticals						
Framing – Verticals/Columns						
Framing – Diagonals						
Crossover Tunnel – Plating						
Hull Openings						
Protective Coating						
Wings						
Inner Shells – Plating						
Inner Shells – Stiffeners						
Inner Shells – Girders						
Outer Shells – Plating						
Outer Shells – Stiffeners						
Outer Shells – Girders						
Trans. End Bulkheads – Plating						
Trans. End Bulkheads – Horizontals						
Trans. End Bulkheads – Verticals						
Trans. Bulkheads – Plating						
Trans. Bulkheads – Horizontals						
Trans. Bulkheads – Verticals						
Framing – Horizontals						
Framing – Diagonals						
Crossover Tunnel – Plating						
Sally Ports – Plating						
Safety Deck – Plating						
Safety Deck – Stiffeners						
Safety Deck – Girders						
Top Deck – Plating						
Top Deck – Stiffeners						
Top Deck – Girders						
Reinforcement for Attachments						

continues

(Continued)

Items Inspected	Condition					Remarks
	U	M	NA	NI	S	
Hull Openings						
Protective Coating						
Welds, Rivets, Bolts						
Exterior Portions						
Pontoon Deck						
Pontoon Side Shells						
Wing Shells – Inner and Outer						
Access Hatches						
Reinforcements for Attachments						
Attachment Fastenings						
Fendering						
Dock Connections						
Shear Connections						
Alignment Connections						
Rub Plating						
Protective Coating						
Aprons/Outriggers						
Deck Plating						
Framing						
Dock Connections						
Planking						
Communication Bridge						
Decking/Grating						
Framing						
Dock Connections						
Blocking						
Keel Blocks						
Side/Bilge Blocks						
Hauling Blocks						
Miscellaneous						
Mooring Spuds and Connections						
Mooring Chains and Anchors						
Cleats						
Bollards						
Chocks						

continues

(Continued)

Items Inspected	Condition					Remarks
	U	M	NA	NI	S	
Gratings						
Platforms/Handrails						
Watertight Doors, Hatches, Manholes						
Access for Operations and Security						
Draftboards						
Stairways/Ladders						
Ship Grounding Straps						
Welding Machine Grounds						
Concrete Hull Structure						
Pontoon Slab – Above Waterline						
Pontoon Slab – Below Waterline						
Wing Wall – Side Slabs						
Wing Wall – End Slabs						
Trans. WT Bulkhead Slabs						
Compartment Slabs						
Storage Tank Slabs						
Access Trunk Slabs						
Pontoon Framing						
Wing Wall Framing						
Storage Tank Framing						
Access Trunk Framing						
Compartment Framing						
Top Decks – Slab						
Top Decks – Attachment Fastenings						
Top Decks – Crane Rail and Supports						
Top Decks – Covering/Painting						
Safety Decks – Slabs						
Safety Decks – Attachment Fastenings						
Safety Decks – Covering/Painting						
Pontoon Deck – Slab						
Pontoon Deck – Sonar Pit Slab						
Pontoon Deck – Attachment Fastenings						
Pontoon Deck – Hatch Frames						
Pontoon Deck – Rub Plating						

continues

(Continued)

Items Inspected	Condition					Remarks
	U	M	NA	NI	S	
Pontoon Sides – Attachment Fastenings						
Sides – Fill & Discharge Sleeves						
Pontoon Sides – SW Sleeves						
Pontoon Sides – Hawsepipe Fastenings						
Pontoon Sides – Ant. Mast Fastenings						
Pontoon Sides – Fenders						
Miscellaneous WT Sleeves						
Blocking						
Keel Blocks						
Side/Bilge Blocks						
Hauling Blocks						
Miscellaneous						
Mooring Spuds and Connections						
Mooring Chains and Anchors						
Cleats						
Bollards						
Chocks						
Gratings						
Platforms/Handrails						
Watertight Doors, Hatches, Manholes						
Access for Operations and Security						
Draftboards						
Stairways/Ladders						

U Unsatisfactory
M Marginal
NA Not Applicable
NI Not Inspected
S Satisfactory

Inspected By: ________

Of Company: ________

Date of Inspection: ________

APPENDIX A2 CERTIFICATION INSPECTION CHECKLIST—GRAVING DOCKS

Facility Name: ___________

Dry Dock Designation: ___________

Items Inspected	Condition U	M	NA	NI	S	Remarks
Dock Structure						
Coping						
Walls						
Galleries						
Altars						
Service Tunnels						
Floor						
Apron						
Caisson Seats						
Drainage Culverts						
Drainage Tunnels						
Flooding Tunnels						
Discharge Tunnels						
General Appearance						
Pressure Relief System						
Dock Fittings						
Cleats						
Bollards						
Chocks						
Gratings						
Platforms/Handrails						
Draft Gauges						
Dock Blocking						
Keel Blocks						
Side/Bilge Blocks						
Hauling Blocks						
Sluice Gates						
Sluice Gate Leaf						
Sluice Gate Guides						
Sluice Gate Operator						
Sluice Gate Controller						
Hydraulic Operating Gear						
Lubrication System						

continues

(Continued)

Items Inspected	Condition					Remarks
	U	M	NA	NI	S	
Preservation						
Check Valves						
Valve Leaves						
Swing Mechanism						
Non-slam Mechanism						
Lubrication System						
Preservation						
Stop Logs or Gates						
Guide Slot						
Log or Gate Body						
Hoisting Equipment						
Intake Screens and Trash Racks						
Preservation						
Fire Protection System						
Fire Pump						
Fire Main						
Fire Stations – Hoses, Nozzles, etc.						
CO_2 Extinguishers						
Dry Chemical Extinguishers						
Miscellaneous						
Access for Operations and Security						
Lighting for Operations and Security						
Capstans						
Dock Mechanical & Electrical						
Capstans						
Capstan Motors						
Capstan Motor Controllers						
Main Dewatering Pumps						
Pump Motors						
Pump Motor Controllers						
Valves & Valve Operators						
Trimming Pumps						
Pump Motors						
Pump Motor Controllers						
Discharge Valves and Valve Operators						

continues

(Continued)

	Condition					
Items Inspected	U	M	NA	NI	S	Remarks
Flood Valves and Valve Operators						
Equalizing Valves and Valve Operators						
Vent Blower						
Blower Motor						
Motor Operator						
Power Leads (Portable)						
Transformers						
Control Panel						
Alarms						
Closure Gate Structure						
Shell Plating						
Deck Plating						
Top Deck Covering						
Bulkhead Plating						
Structural Framing						
Fenders						
Backing for Seals						
Seals						
Fixed Ballast						
Exterior Preservation						
Interior Preservation						
Ballast Compartment Preservation						
General Condition						
Closure Gate Fittings						
Hatches						
Cleats						
Bollards						
Chocks						
Gratings						
Platforms/Handrails						
Air Ports						
Compressed Air Piping						
Compressed Air Control Valves						
Inclinometers						
Water Level Indicators						

continues

(Continued)

Items Inspected	Condition					Remarks
	U	M	NA	NI	S	
Utility Services						
Compressed Air Piping and Valves						
Power Distribution Systems						
Communication Systems						
Ventilation Systems						
Dry Chemical Extinguishers						
CO_2 Extinguishers						
Pumping & Dewatering System – Mechanical						
Pumps						
Pump Motors						
Motor Controllers						
Shafts & Couplings						
Guide Bearings						
Impellers						
Wearing Rings						
Pump Casings						
Packing Glands						
Lubrication System						
Flanges & Gaskets						
Drainage Pumps						
Sump Pumps						
Pump Motors						
Motor Controllers						
Suction Valves and Operators						
Discharge Valves and Operators						
Check Valves and Operators						
Valve Controllers						
Air Blower Motor						
Blower Controller						
Lubrication System						
Preservation						
Pumping & Dewatering System – Electrical						
Control Panel						
Selector Switches						
Pushbutton Controls						

continues

(Continued)

Items Inspected	Condition					Remarks
	U	M	NA	NI	S	
Indicator Lights						
Feeder Controls						
Valve Indicators						
Water Level Indicators						
Trouble Indicators and Alarms						
Transfer Switches						
Reset Buttons						
Relays						
Voltmeters						
Ammeters						
Temperature Indicators						
Transformers						
Circuit Breakers						
Pump House Structure						
Roof						
Walls						
Floor						
Pump Pits						
Suction Chambers						
Suction Bells						
General Preservation						
Hatches						
Gratings						
Platforms/Handrails						

U Unsatisfactory
M Marginal
NA Not Applicable
NI Not Inspected
S Satisfactory

Inspected By: ________

Of Company: ________

Date of Inspection: ________

APPENDIX A3 CERTIFICATION INSPECTION CHECKLIST—MARINE RAILWAYS

Facility Name: ___________

Dry Dock Designation: ___________

Items Inspected	Condition					Remarks
	U	M	NA	NI	S	
Cradle						
General Condition						
Decking						
Cradle Beams						
Uprights and Catwalks						
Columns						
Runners						
Hauling Connection						
Hauling Chain Equalizer Sheave						
Backing Chain Equalizer Sheave						
Wheel Bearing Supports						
Low Logs						
Groundways & Rails						
Settlement of Tracks						
Condition of Piles						
Condition of Stringers						
Condition of Cross Bracings						
Track Plates and Fasteners						
Rails and Fasteners						
Condition of Chain Guides						
Preservation						
Rail Alignment						Line and Grade
Mud and Silt Condition						
Wheels						
Wheel Bearings						
Rollers						
Roller Bushings						
Roller Frames						
Spacer Blocks						
Fittings						
Cleats						
Ringbolts						
Draftboards						

continues

(Continued)

Items Inspected	Condition					Remarks
	U	M	NA	NI	S	
Blocking						
Keel Blocks						
Side/Bilge Blocks						
Hauling Blocks						
Hauling Chains						
Locking Pawls						
Chains & Sheaves						
Inhaul Chains/Cable						Gauge Chains
Outhaul Chains						Gauge Chains
Inhaul Sheaves						
Outhaul Sheaves						
Chain Connecting Links						
Sheave Fasteners						
Chain Slack and Fit						
Preservation						
Hauling Machinery						
Gearing, Speed Reducer						
Shafting						
Bearings						
Sprockets, Chain Wheels						
Cable Drum						
Frame						
Anchor Bolts						
Electric Brakes						
Hand Brakes						
Locking Pawls						
Clutch						
Safety Guards						
Lubrication						
Preservation						
Electric Motor						
Diesel/Gas Engine						
Steam/Compressed Air Drives						
Controller						
Speed Limit Device						

continues

(Continued)

Items Inspected	Condition					Remarks
	U	M	NA	NI	S	
Electrical						
Control Board						
Switches						
Safety Devices and Alarms						
Communication System						
Public Address System						
Radios						
Fire Alarm						
Mechanical						
Bilge Block Hauling Equipment						
Warping Winches						
Head Capstan						
Fire Protection System						
Fire Pump						
Fire Main						
Fire Stations – Hoses, Nozzles, etc.						
CO_2 Extinguishers						
Dry Chemical Extinguishers						
Miscellaneous						
Access for Operations and Security						
Lighting for Operations and Security						
General Condition of Winch House						

U Unsatisfactory
M Marginal
NA Not Applicable
NI Not Inspected
S Satisfactory

Inspected By: ________

Of Company: ________

Date of Inspection: ________

APPENDIX A4 CERTIFICATION INSPECTION CHECKLIST—VERTICAL LIFTS

Facility Name: ___________

Dry Dock Designation: ___________

Items Inspected	Condition					Remarks
	U	M	NA	NI	S	
Hoists						
Motors						
Gears						
Brakes						
Wire Ropes						
Bearings						
Drums						
Foundation Platforms						
Anchorage						
Piles						
Pawls						
Lubrication						
Preservation						
Wiring						
Platform						
Main Transverse Beams						
Secondary Transverse Beams						
Longitudinal Beams						
Stiffeners						
Decking						
Sheaves						
Bearings						
Sheaves Housing						
Lubrication						
Tracks						
Preservation						
Pins						
Cradles						
Main Transverse Beams						
Secondary Transverse Beams						
Longitudinal Beams						
Stiffeners						
Wheels/Rollers/Roller Plates						

continues

(Continued)

Items Inspected	Condition					Remarks
	U	M	NA	NI	S	
Roller Spindles/Wheel Axles						
Block Bearers						
Preservation						
Compressed Air						
Compressors/Receivers						
Air Dryers						
Valves and Piping						
Controls						
Controls						
Circuit Breakers						
Motor Starters						
Disconnect Switches						
Limit Switches						
Load Sensors						
Load Indicators						
Cams						
Ammeters						
Voltmeters						
Push Buttons						
Drum Controllers						
Electrical Interlocks						
Mechanical Interlocks						
Indicating Lights						
Relays						
Alarms						
Transfer System						
Tracks						
Hauling Equipment						
Load Bearers						
Cradles						
Tracks						
Blocking						
Keel Blocks						
Side/Bilge Blocks						
Hauling Blocks						

continues

(Continued)

Items Inspected	Condition					Remarks
	U	M	NA	NI	S	
Communication System						
Public Address System						
Radios						
Fire Alarm						
Power System						
Main Power Supply						
Backup Power Supply						
Power Distribution System						
Fire Protection System						
Fire Pump						
Fire Main						
Fire Stations – Hoses, Nozzles, etc.						
CO_2 Extinguishers						
Dry Chemical Extinguishers						

U Unsatisfactory
M Marginal
NA Not Applicable
NI Not Inspected
S Satisfactory

Inspected By: _________

Of Company: _________

Date of Inspection: _________

APPENDIX B
RECOMMENDED PREVENTIVE MAINTENANCE PERIODICITIES

SAMPLE FLOATING DRY DOCK MAINTENANCE TASKS

This appendix provides a minimum preventive maintenance program for floating dry docks. This is meant as a starting point. If the operator has historical data to show that an alternative program is providing acceptable results, that program should be continued. Furthermore, this preventive maintenance program is a traditional time-based program. It is suggested that metrics be obtained and studied to develop a condition-based maintenance program. A properly designed and implemented condition-based maintenance program is expected to be more efficient and cost-effective.

5-Year PM

- Mooring Arm Components: Disassemble, inspect, replace/rebuild worn parts as deemed necessary, and lubricate.
- Capstan Motors: Inspect and lubricate.
- Water Depth: Sound water depth under cradle and approaches. Dredge as required.

3-Year PM

- Mooring Arm Components: Blast and recoat as needed.
- Pumps and Gearboxes: Replace oil.

Biennial PM

- Air Compressors: Replace oil.
- Motors: Replace motor-bearing oil.

Annual PM

- Structure: Inspect all structural members.
- Valve Operators: Inspect. Lubricate/repair as required.
- Capstans: Lubricate.
- Motors, Drive Gearboxes, and Drive Chains: Lubricate.
- Mooring Arm Components: Inspect. Take micrometer measurements of connection pin to identify any extraordinary wear.
- Oily Waste System Piping: Pneumatic pressure test.
- Fuel Oil System Piping: Pressure test.
- Reduced Pressure Principal Backflow Preventers, Pressure Vacuum Breaker Devices, and Double Check Valve Assemblies: Test.
- Piping, Valves, Piping Supports, and Pipe Identification: Inspect.
- Cathodic Protection: Inspect zinc/aluminum anodes and replace as necessary.

Semiannual PM

- Phone Systems: Test operations.
- Mooring Arm Components: Inspect and lubricate.
- Ballast Tank Hatches: Inspect and lubricate.
- Freeze Protection Valves: Maintain.
- Hydraulic Fluids: Sample.

Quarterly PM

- High Water Alarms: Inspect/test.
- Equipment with Oil and/or Hydraulic Fluid: Inspect/add as needed.
- Capstans: Lubricate.
- Pumps and Pump Discharge Flap Valves: Lubricate.
- Air Compressors: Inspect oil level and belts and drain moisture from tanks.
- Level Sensor Tubes: Flush.
- Fire Pumps and Sprinkler Pumps: Replace snubbers.

2-Month PM

- Mooring Arm Components: Lubricate slider rail.

Monthly PM

- Egress Lighting Equipment: Inspect and test.
- Mooring Arm T-Rails: Lubricate.
- Freeze Protection Valves: Test.
- Fire Pumps: Operate.

Weekly PM

- Ballast System Valves and Pumps: Operate.

SAMPLE GRAVING DOCK MAINTENANCE TASKS

This appendix provides a minimum preventive maintenance program for graving docks. This is meant as a starting point. If the operator has historical data to show that an alternative program is providing acceptable results, that program should be continued. Furthermore, this preventive maintenance program is a traditional time-based program. It is suggested that metrics be obtained and studied to develop a condition-based maintenance program. A properly designed and implemented condition-based maintenance program is expected to be more efficient and cost-effective.

PUMPWELL
Biennial PM

Drainage Pump: Change motor coolant. Remove impeller for inspection and measurement.
Main Dewatering Pump: Sample/change oil as necessary.

Annual PM

Structure: Inspect all structural members.
Wetwell Recorder: Clean and inspect.
Pressure Switches for Drainage Pump Control: Clean and inspect.

Pumpwell Air Compressor: Calibrate pressure gauge and set/check pressure switch.

Motors for Vent Fans, Sewage Pump, Sump Pumps, and Air Compressor: Megger motor, megger line, clean/inspect motor controllers, inspect windings, and check wiring connections.

Main Dewatering Pump (MDP) Motor: Clean/inspect motor windings, slip rings, brush rigging, resistance grid, and contactor. Megger rotor, stator, and resistance grid.

MDP Discharge Valve and Sluice Gate Motors: Clean/inspect motor and controller. Inspect wiring insulation. Megger motor windings phase to ground.

Sluice Gates: Inspect gate, frame, thimble. Check fasteners for tightness. Check locking bolts/keys. Check clearances on frame.

Drainage Pump Motors: Clean/inspect motor and controller. Inspect wiring insulation. Check connections for tightness. Megger motor windings phase to ground. Change oil. Perform capacity checks. Record motor run times.

Drainage Pump Valve Actuators: Clean/inspect motor and controller. Inspect wiring insulation. Check connections for tightness.

Cathodic Protection: Inspect zinc/aluminum anodes and replace as necessary.

Semiannual PM

Pumpwell Vent Fans: Clean/inspect/preserve housing, motor, and fan blades. Inspect/replace/adjust fan belts. Check oil and lubricate.

Pumpwell Air Supply Filters: Inspect/replace.

Pumpwell Air Compressor: Inspect housing, tank, and motor. Drain condensate. Check belts. Check oil levels.

Pumpwell Sump and Sewage Pumps: Inspect/clean/preserve. Exercise Discharge Valves. Lubricate.

Main Dewatering Pumps: Clean packing gland and fasteners. Inspect pump supports. Exercise vent valves.

MDP Line Shaft Bearings: Inspect/preserve. Check oil levels.

MDP Discharge Valves: Inspect/lubricate valve operators. Inspect valve packing. Clean packing gland and fasteners. Inspect/exercise valves.

Sluice Gates and Operators: Clean/inspect/grease stem threads. Inspect packing. Lubricate/preserve operator. Exercise gate.

MDP Check Valves: Check dashpot oil. Clean/preserve, as necessary. Inspect dashpot packing. Lubricate trunnion bearings.

Suction Chamber Anodes: Inspect/replace, as necessary.

Drainage Pumps: Operate pumps. Inspect for leaks. Clean/preserve.

Drainage Pump Motors: Inspect/clean/preserve.

Drainage Pump Valves and Operators: Inspect/clean/preserve valves, packing and fasteners. Lubricate operators and steady bearings. Exercise operators/valves.

Drainage Pump Check Valves: Clean/preserve.

Quarterly PM

Secondary Transducer Battery Load Test: Test battery capacity under load.

Monthly PM

Pumpwell Sump Pumps: Inspect and operate floats.

High Water Level Alarm Tests: Test dry pit sump pump high level alarm. Test wetwell and dry dock transducer high water level alarms (both secondary and primary transducers). Test backup bubbler system high water alarm points.

CAISSON

Annual PM

Caisson Valves: Inspect/exercise all caisson valves.

Caisson Tank Level Indicators (TLIs): Inspect TLI panel and associated wiring, lights, and switches. Test high water alarm setpoints and alarm horns.

Cathodic Protection: Inspect zinc/aluminum anodes and replace as necessary.

Semiannual PM

Caisson Valves and Motors: Inspect motors, controllers, panel bussing, and receptacles. Megger motor windings. Megger lines back to caisson plug. Lubricate all valves and pumps. Inspect pump packing. Preserve as necessary.

Quarterly PM

Caisson Valves and Pumps: Inspect/exercise all valves and pumps. Add oil to motors and lubricate via grease lines.

DRY DOCK

Every 5 Years

Water Depth: Sound water depth in approaches. Dredge as required.

Biennial PM

Dry Dock Suction Grates: Inspect suction grates by sluice gates.

Annual PM

Tunnel Gates and Operators: Clean/inspect/grease stem threads. Inspect packing. Lubricate/preserve operator. Exercise gate. Inspect gate, frame, and thimble. Check fasteners for tightness. Check locking bolts/keys. Check clearances on frame.

Semiannual PM

Tunnel Gate Operator Motors: Inspect motor for signs of overheating or damaged wiring. Inspect/clean motor controller. Megger motor phase to ground. Megger line from controller to breaker.

Tunnel Gate and Operator: Check grease levels in operator. Clean/preserve operator. Exercise gate. Exercise inner seat drain valve. Inspect mooring cleat, chocks, and bollards. Inspect/exercise sump pump and drain valve.

Monthly PM

Dry Dock Tunnel Gates: Inspect/exercise sump pump and floor drain.

CAPSTANS

Annual PM

Capstans: Inspect motor and machinery foundations. Inspect structural fasteners. Inspect motor case and gearbox for cracks. Inspect for worn defective or misaligned couplings, bushings, shafts, pins, gears, and drums. Check lubricant levels. Check brake pad thickness. Operate capstans and observe proper function of motors, controls, brakes, and gears.

Semiannual PM

Capstans: Inspect controller. Inspect cables, wiring, and pendant controller. Megger capstan motor. Lubricate motor if necessary. Inspect capstan brakes. Lubricate upper capstan shaft.

Monthly PM

Capstans: Inspect/preserve/clean capstans and capstan pits. Check oil levels. Exercise capstans in both speeds in both directions.

SAMPLE MARINE RAILWAY MAINTENANCE TASKS

This appendix provides a minimum preventive maintenance program for marine railways. This is meant as a starting point. If the operator has historical data to show that an alternative program is providing acceptable results, that program should be continued. Furthermore, this preventive maintenance program is a traditional time-based program. It is suggested that metrics be obtained and studied to develop a condition-based maintenance program. A properly designed and implemented condition-based maintenance program is expected to be more efficient and cost-effective.

Every 5 Years
Hauling Chain: Gauge inhaul and outhaul chain.
Water Depth: Sound water depth under cradle and approaches. Dredge as required.

Every 2 Years
Track: Check track gauge, line, and grade.

Annually
Structure: Inspect all structural members.
Power and Control Systems: Inspect mechanical and electrical components.
Cathodic Protection: Inspect zinc/aluminum anodes and replace, as necessary.

Semiannual
Inspect and tightened hauling chains as required.

Prior to Each Block Build
Cleanliness: Clean railway and associated work area.
Structure: Inspect cradle structure and fasteners.
Fixed Blocks: Inspect keel blocks and fixed side blocks.
Sliding Blocks: Inspect sliding side blocks, sliding, locking, and hauling hardware.

Prior to Each Cradle Movement
Tracks: Wash rollers and tracks.
Bearings: Grease bearings.
Downhaul System: Inspect downhaul chain/cable, check attachment point, and adjust tension.
Blocks: Check all block straps and fasteners.
Cleanliness: Clean railway and associated work area. Remove foreign material from cradle.
Brake: Inspect brake.

SAMPLE VERTICAL LIFT MAINTENANCE TASKS

This appendix provides a minimum preventive maintenance program for vertical lifts. This is meant as a starting point. If the operator has historical data to show that an alternative program is providing acceptable results, that program should be continued. Furthermore, this preventive maintenance program is a traditional time-based program. It is suggested that metrics be obtained and studied to develop a condition-based maintenance program. A properly designed and implemented condition-based maintenance program is expected to be more efficient and cost-effective.

Every 10 Years
Wire Hoist Ropes: Replace all.
Hoist Motors: Strip, clean, and examine stator windings. Renew bearings and reassemble.
Lift Platform: Ultrasonically inspect butt welds. Visually inspect structure for corrosion loss.

Every 8 Years
Upper Sheave Bearings: Strip and examine one set.

Every 5 Years
Water Depth: Sound water depth under cradle and approaches. Dredge as required.

Every 2 Years
Protective Coatings: Touch up hoist and cradle paint as necessary.
Hoist Motors: Inspect flexible couplings.

Annually
Structure: Inspect all structural members.
Hoist Pawls: Inspect for proper engagement.
Gear Box Oil: Test and replace, as necessary.
Fasteners: Check all bolts for tightness, particularly those on rope clamps and drum open gear.
Solenoid Air Gap: Check.
Brakes: Dismantle and check the friction and stationary discs. Check torque.
Hoist Motors: Test insulation resistance.
Lower Sheaves: Strip and examine one set.
Cathodic Protection: Inspect zinc/aluminum anodes and replace as necessary.
Control and Protection: Check.

Every 6 Months
Wire Rope Assemblies: Inspect.
Upper Sheaves: Lubricate bearings.
Limit Switches: Grease actuator rod.
Pawl Shafts: Lubricate.
Brakes: Clean, inspect, check torque and solenoid air gap.

Every 3 Months
Gear Boxes: Check oil level.

Monthly
Inspect and test all limit switches.

Weekly
Wire Ropes: Inspect and lubricate as required.
Open Gears: Inspect and lubricate as required.
Lower Sheave Bearings: Grease.
Platform Decking: Inspect.
Transfer System: Inspect and lubricate cradle wheel bearings.

APPENDIX C
DRYDOCKING CHECKLIST

Because of the criticality of dry dock operations, a checklist shall be used to ensure that all steps are taken to facilitate a successful outcome. This checklist shall be used as a guideline to develop site specific checklists for each facility. Some of the items may not apply to all drydocking or undocking operations. Other items not listed in this checklist that are unique facility characteristics shall be included in the final checklists.

PRE-DRYDOCKING CONFERENCE CHECKLIST

ITEM	SAT	UNSAT
DOCUMENTATION TO BE PROVIDED		
Current dock certification		
Operating practices, safety requirements, and yard security plans		
Docking calculations		
Blocking arrangement (if different from docking plan)		
Docking procedure		
FACILITY SAFETY EQUIPMENT		
Fire alarm locations		
Emergency power		
Emergency ballast/dewatering pumps		
REVIEW		
Flooding and pumping plan for the dry dock (allowable trim and deflection)		
Specific list, trim and drafts of the vessel during docking (grounding, when blocks are hauled) Critical Draft___________		
GM of ship dock system all phases (floating DD only—not less than 5 ft, except on docks of greater than 10,000 LT capacity)		
Block loading—Trapezoidal, Knuckle		
Any special precautions or actions characteristic to the docking facility, the docked vessel, or a combination		
High/low water, currents, weather		
Communications plan		
Tug plan		
Vessel entry plan (line handlers, fenders)		
Vessel clearance above keel blocks, side blocks, and other potential obstructions		
Docking position		
Procedure for positioning vessel in dock		
When to secure ship's power		
Use of divers		
Arrange time for block inspection		
Time and date of drydocking __________________		
VESSEL CONDITION		
Verify vessel load condition (tanks, drafts, displacement)		
All equipment retracted		
Verify temporary services/hookups		
Drafts: FWD________, MID_________, AFT________		

PRE-DRYDOCKING DOCK INSPECTION

ITEM	SAT	UNSAT
FOUNDATION BLOCK—TIMBER		
Check timber for excessive crushing, warping, cracking, rot, and degraded material		
Note amount of wear from spiking and dogging		
Evaluate the condition of the interfacing between blocks in the stack		
Note condition of the fasteners in the blocks		
Note arrangements for preventing tripping and floating of blocks		
FOUNDATION BLOCK—CONCRETE		
Structural damage because of overloads		
Corrosion of steel reinforcement		
Concrete for cracking, spalling, and exposed rebar		
FOUNDATION BLOCK—STEEL		
Evaluate the loss of steel owing to corrosion		
Look for cracks in welds		
Deformed structure		
BLOCKS—GENERAL		
Soft caps minimum thickness 2 in. and no crush		
Spacing and location per blocking arrangement (+/– 1/2 in. transversely +/– 1 in.; longitudinally, +/– 1/4 in. height)		
Perform hauling block system operational check		
KEEL BLOCKS		
Sight keel block line for alignment and fit		
Keel block height meets requirement		
Keel profile applied to keel block offsets		
SIDE BLOCKS		
Sight side block line for alignment and fit		
Side blocks are required dimensions		
Side block construction (Normal force passes through middle 1/3 of all blocks, no gaps, cribbing if over 6 ft)		
MISCELLANEOUS		
Crane clearance		
Check overhead interferences and clearances		
Depth of water (tide dependent)		
Condition of the working floor for debris, unevenness, and so forth		
Note mooring system, possibility of streaming		
Note draft/trim devices in use		
Condition of fendering		
Condition of lifting straps		
Perform ship handling system operational check		

DURING AND POST DRYDOCKING INSPECTION

ITEM	SAT	UNSAT
DURING DRYDOCKING OPERATION		
Time and date when bow crosses sill (or opening) ___________, ______________		
Vessel came in smoothly. Could it have hit any underwater obstacles?		
Position of the vessel is correct		
Correct draft of dock when vessel grounds		
Correct drafts of dock and vessel when vessel is landed		
Check for vessel list and alignment		
Correct draft of vessel when side blocks are hauled		
All side blocks were hauled fully		
Draft of vessel when setting down on preset side blocks		
Keel centered on keel blocks		
Trim and docking plan being followed		
POST DRYDOCKING OPERATION		
Proper contact area (wedges may be required). If inadequate area refloat (less than 80%)		
Drafts of dock (FWD_______, MID_______ , AFT________)		
Does dock have a hog or sag?		
Are any blocks hitting appendages?		
Any appendages not shown on docking plan or in wrong location?		
Excessive crush of blocks? Location: ______________________________		
Verify correct position of vessel on blocks		
Ensure side haul blocks are locked in position		
Damage to vessel (describe)		

PRE-UNDOCKING CONFERENCE CHECKLIST

ITEM	SAT	UNSAT
DOCUMENTATION TO BE PROVIDED		
Recorded weight shifts during availability		
Undocking calculations		
Undocking procedure		
UNDOCKING REPORT		
Transducers uncovered		
Zincs uncovered and free of paint		
Shaft rope guard and fairwaters in place		
Hull opening blanks and plugs removed		
Sea chest strainers are bolted in place and lockwired		
Sea valves and waster pieces are properly installed and are in the closed position		
All underwater body work has been completed		
Dock is free of all debris and blasting material		
REVIEW		
Flooding and pumping plan for the dry dock. (allowable trim and deflection)		
Specific list, trim, and drafts of the vessel during undocking. (when side blocks are hauled)		
GM of ship dock system all phases (floating DD only—not less than 5 ft, except on docks of greater than 10,000 LT capacity)		
High/low water, currents, weather		
Communications plan		

continues

PRE-UNDOCKING CONFERENCE CHECKLIST. *(Continued)*

ITEM	SAT	UNSAT
Tug plan		
Perform ship handling system operational check		
Temporary services disconnection		
Vessel exit plan (line handlers, fenders)		
Vessel clearance above keel blocks, side blocks, and other potential obstructions		
Pier location and temporary services hookup		
Where personnel will be stationed (all hull openings that were worked on)		
Procedure if immediate redocking is required		
Is ballast required for undocking?		
Time and date of undocking ________, _________		
VESSEL CONDITION		
Verify vessel load condition (tanks, drafts, displacement). Perform tank sounding within 12 hours of undocking.		

UNDOCKING INSPECTION

ITEM	SAT	UNSAT
DURING UNDOCKING OPERATION		
All equipment retracted		
Verify temporary services/disconnection		
Personnel at hull openings		
Stopped at correct draft for hauling side blocks		
Hauled ALL side blocks FULLY		
Detection of any leaks		
Vessel exited smoothly. Could it have hit any underwater obstacles?		
Time and date when bow crosses sill _____________		
Drafts: FWD________, MID_________, AFT________		
Damage: (describe)		

INDEX

Page numbers followed by *f* or *t* indicate figures or tables.